Coleoptera: Elmidae and Protelmidae

World Catalogue of Insects

VOLUME 14

The titles published in this series are listed at *brill.com/wci*

Ancyronyx acaroides (*left*) *and* A. malickyi (*middle and right*) *on submerged wood in a Sumatran lowland river. Painting by W. Zelenka (†), ca. 2000.*

Coleoptera: Elmidae and Protelmidae

By

Manfred A. Jäch
Ján Kodada
Michaela Brojer
William D. Shepard
Fedor Čiampor, Jr.

BRILL

LEIDEN | BOSTON

Cover illustrations: Habitus illustrations (from above): Cuspidevia, Macronychus, Potamophilus, Stenelmis, Troglelmis. Paintings by W. Zelenka (†).

Library of Congress Control Number: 2006356329

Typeface for the Latin, Greek, and Cyrillic scripts: "Brill". See and download: brill.com/brill-typeface.

ISSN 1398-8700
ISBN 978-90-04-29176-8 (hardback)
ISBN 978-90-04-29177-5 (e-book)

This book is printed on acid-free paper and produced in a sustainable manner.

Contents

Acknowledgements

The following persons are sincerely thanked for various advice, for providing literature and/or distributional data, or for sending specimens: V.I. Alekseev (Kaliningrad, Russia), M.A. Alonso-Zarazaga (Madrid, Spain), R.B. Angus (London, U.K.), M. Barcley (London, U.K.), E. Bauernfeind (Vienna, Austria), R. Booth (London, U.K.), D.S. Boukal (České Budějovice, Czech Rep.), S. Bouzid (Annaba, Algeria), W. Brunnbauer (Vienna, Austria), B. Bruvo Mađarić (Zagreb, Croatia), P. Buzinsky (Lublin, Poland), E. de Coninck (Tervuren, Belgium), G.N. Foster (Ayr, U.K.), N. Ferreira (Rio de Janeiro, Brazil), M. Fikáček (Prague, Czech Rep.), H. Freitag (Manila, Philippines), J. Garrido (Vigo, Spain), M. Gonzáles-Córdoba (Cali, Colombia), M. Hess (Munich, Germany), H. Huijbregts (Leiden, Netherlands), L. Ji (Shenyang, China), S.-W. Jung (Seoul, South Korea), Y. Kamite (Nagoya, Japan), A. Komarek (Vienna, Austria), A. Kovalev (St. Petersburg, Russia), J. Krikken (Leiden, Netherlands), M. Madl (Frauenkirchen, Austria), A. Mantilleri (Paris, France), M. Maruyama (Fukuoka, Japan), O. Merkl (Budapest, Hungary), V. Mičetić Stanković (Zagreb, Croatia), A. Millán (Murcia, Spain), H. Nasserzadeh (Tehran, Iran), N.B. Nikitskyi (Moscow, Russia), A.N. Nilsson (Umeå, Sweden), A. Prokin (Voronezh, Russia), M. Przewoczny (Poznań, Poland), I. Ribera (Barcelona, Spain), W. Sondermann (Bogotá, Colombia), M. Tavano (Genova, Italy), C. Taylor (London, U.K.), D. Telnov (Riga, Latvia) and H. Yoshitomi (Matsuyama, Japan).

P.J. Spangler (†) and H.P. Brown (†) are thanked for sending an unpublished checklist to W.D. Shepard many years ago.

All habitus illustrations (book cover and frontispiece) were kindly provided by W. Zelenka (†).

This study was partly supported by the Slovak Research and Development Agency: Project No. APVV-0213-10 and VEGA: Project No. 1/0395/14.

World Catalogue of Insects. Vol. 14. Coleoptera: Elmidae, Protelmidae

M. A. Jäch & M. Brojer
Naturhistorisches Museum Wien, Burgring 7, A – 1010 Wien, Austria (manfred.jaech@nhm-wien.ac.at (corresponding author), michaela.brojer @nhm-wien.ac.at)

J. Kodada
Faculty of Natural Science, Comenius University in Bratislava, Mlynská dolina Ilkovičova 6, 842 15 Bratislava 4, Slovakia (kodada@fns.uniba.sk)

W.D. Shepard
Essig Museum of Entomology, 1101 Valley Life Sciences Bldg., #4780, University of California, Berkeley, California 94720 USA (william.shepard@csus.edu)

F. Čiampor, Jr.
Zoology Lab, Institute of Botany, Slovak Academy of Sciences, Dúbravská cesta 9, SK – 845 23 Bratislava, Slovakia (f.ciampor@savba.sk)

Introduction

Elmidae or Riffle Beetles are a moderately large cosmopolitan family of True Water Beetles (see Jäch & Balke 2008). Adults and larvae are usually considered to be aquatic, however, adults of several genera, e.g. *Potamophilus* Germar, 1811 are often encountered a little above the water line or in spray zones of waterfalls and cascades. Members of this family are generally living in lotic habitats, few species may be found in lakes or ponds and a few species are recorded from subterranean waters. Strictly terrestrial representatives, which do exist in other typical water beetle families (e.g. Dytiscidae, Hydrophilidae, Hydraenidae, Dryopidae) are completely unknown among elmids. Pupation takes place above the water line.

Elmidae are often used in ecological running water monitoring programs (Moog & Jäch 2003). In the past, elmids have even been part of the diet of man. *Austrelmis condimentarius* was used as seasoning for food in South America. This species was reported to have considerable commercial value (Philippi 1864).

The morphology of Elmidae was described in detail by Kodada & Jäch (2005) (second edition in press).

Elmidae belong to Polyphaga and are currently placed in Elateriformia. Within Elateriformia, they are usually attributed to Byrrhoidea (Kodada & Jäch 2005, Bouchard et al. 2011, Kundrata et al. 2013, Bocak et al. 2014), or Dryopoidea (Crowson 1955, 1981, Lawrence & Newton 1982, Hunt et al. 2007).

Protelmidae, hitherto regarded as a tribe of Elminae, are here elevated to family rank, because of strongly deviating external and genitalic features. Adults of Protelmidae are aquatic (three genera live in riffles of running water, one genus is troglobiontic). Protelmid larvae are still unknown.

Among the "great six" water beetle families (i.e. families with more than thousand species, see Jäch & Balke 2008), only the three largest ones (Dytiscidae, Hydraenidae, Hydrophilidae) have been covered by world catalogues recently.

The last world catalogue of Elmidae was published more than one hundred years ago, by Zaitzev (1910) and included 297 species. Elmidae were still regarded as a subfamily of Dryopidae by Zaitzev (1910), although various authors have already treated Elmidae as a distinct family earlier (e.g. Grouvelle 1889b, 1892a, 1896a, d, 1900b, 1906b, c, 1908, Reitter 1886, 1899).

The North American Elmidae were catalogued by Brown (1983), the Palearctic elmids were published by Jäch et al. (2006) (second edition in press), and the Neotropical taxa were compiled by Segura et al. (2013). Catalogues for the Australian, Afrotropical, and the Oriental fauna have never been published, but a check list of the Australian species was provided by Glaister (1999).

A preliminary version of this catalogue was published by Joel K. Hallan in the internet ("Synopsis of the described Coleoptera of the world": http://insects.tamu.edu/research/collection/hallan/test/Arthropoda/Insects/Coleoptera/Family/Coleoptera1.htm), based on a card catalogue by H.P. Brown (1921–2008), authored by M.A. Jäch and J. Kodada. This preliminary version was updated until June 2010. The reason for placing this preliminary version in the internet was simply to provide a widely accessible source of useful information, although it was still very incomplete, especially with regard to distributional data.

Particular Problems

In the course of the compilation of this catalogue we faced numerous nomenclatorial problems. These concerned for instance the correct spelling of taxa names (e.g. decisions between correct and incorrect original spellings (sensu ICZN 1999: Art. 32), or treatment of epithets that are nouns in apposition), treatment of "subsequent spellings" (sensu ICZN 1999: Art. 33), correct identity and spelling of author names, correct publication dates, correct type localities, the availability and correct authority of some very poorly described species published in Zhang & Yang (1995), and finally the unsatisfactory quality of various global online lists.

We were able to check all references listed in this catalogue, except for one: Crotch (1872), who published a nomen nudum (*Limnius sinaiticus*).

Taxa Names

Many of the species group epithets erected by Hinton are definitely nouns in the nominative singular standing in apposition to the generic name (ICZN 1999: Art. 11.9.1.2). They are independent of the grammatical gender of the generic name (ICZN 1999: Art. 31.2.1). Subsequent authors, who did not realize, that these epithet names were in fact nouns, unwarrantedly used their adjectival forms or changed their gender when transferring the species to genera of different sex. Some examples: *Macrelmis aeolus* (Hinton, 1946b) [not *aeolis*], *M. aleus* (Hinton, 1946b) [not *alea*], *M. aristaeus* (Hinton, 1946b) [not *aristea*], *M. celsus* (Hinton, 1946b) [not *celsa*], *M. codrus* (Hinton, 1946b) [not *codris*], *M. isus* (Hinton, 1946b) [not *isis*], *M. saleius* (Hinton, 1946b) [not *saleia*], *M. tereus* (Hinton, 1946b) [not *terea*], *M. tityrus* (Hinton, 1946b) [not *tityra*], *M. zamus* (Hinton, 1946b) [not *zama*].

In case of *Austrelmis condimentarius*, by far the oldest available name in the genus *Austrelmis*, the epithet name can be regarded as an adjective (condimentarius, –a, –um [pertaining to a spice]) or as a noun in apposition (condimentarius, masculinum [spice merchant]), and therefore, according to ICZN (1999: Art. 31.2.2) the original spelling, *condimentarius*, is to be treated as a noun in apposition, with gender ending unchanged.

Joseph Delève described two species of Elmidae after the Austrian scientist Ferdinand Starmühlner (1927–2006): *Aruelmis starmuhlneri* Delève, 1973d (*Ilamelmis*) and *Pseudelmidolia starmuhleri* Delève, 1963d. In both cases the epithet name is incorrectly spelled. According to ICZN (1999: Art. 32.5.1) the epithet names cannot be changed to "starmuehlneri", because there is in the original publication itself, without recourse to any external source of information, no clear evidence of an inadvertant error, such as a lapsus calami or a copyist's or printer's error.

Stenelmis guangxinensis was named after Guangxi Zhuang Autonomous Region (China), but spelled *guangxinensis* throughout the original description (4 ×) although the correct name should be "*guangxiensis*". According to ICZN (1999: Art. 32.5.1) it is not entirely clear, whether the original spelling of this name is incorrect ("lapsus calami") or correct ("incorrect latinization"); in his unpublished thesis, Zhang (1994) spelled it as "*guangxiensis*" (p. 23) and as *guangxinensis* (p. 52). In order to maintain stability we regard the name *guangxinensis* as correct original spelling ("incorrect latinization").

In case of *Cylloepus friburguensis* and *Macrelmis friburguensis*, the decision about correct and incorrect original spelling is somewhat ambiguous according to ICZN (1999: Art. 32). Both species were named after Nova Friburgo in Brazil. Although the correct Latin name of Friburgo is Friburgum (stem: Friburg-), the correct adjective must be "*friburgensis*", not *friburguensis*. However, "use of an inappropriate connecting vowel" is not to be considered as an inadvertant error (ICZN 1999: Art. 32.5.1). In this case, the vowel "u" might be regarded as (an inappropriate) vowel connecting the latinized stem "Friburg-" and the ending "-ensis". Therefore we decided to treat the epithet name *friburguensis* as correct original spelling in both cases. This means, that the grammatically correct name *Macrelmis* "*friburgensis*", used by Segura et al. (2013: 26) has to be regarded as incorrect subsequent spelling.

Subsequent Spellings

Distinction between "unjustified emendation" (available) and "incorrect subsequent spelling" (unvailable) is, in several cases, quite difficult. For instance, *Lophelmis* Zaitzev, 1908 (footnote) is to be regarded as an "incorrect subsequent spelling" of *Lobelmis* Fairmaire, 1898b and therefore is not available (ICZN 1999: Art. 33.3), whereas *Lophelmis* Zaitzev, 1910 is to be regarded as an "unjustified emendation" of *Lobelmis* and therefore is available (ICZN 1999: Art. 33.2.1) and has to be regarded as a synonym of *Lobelmis*.

There are numerous incorrect subsequent spellings found in various internet lists claiming to provide global information on taxon names (see below, under "Internet lists"). Although we listed a few misspellings of genus group names ("*Cylhepus*", "*Pammicrodinodes*", "*Vietelrnis*") in this catalogue, we did not include misspellings of species group names found in these internet lists.

Author Names

Herein, all authors are spelled in the same version in which they were published in the respective original papers. There are several versions for certain authors, e.g. Le Conte (1852, 1869 [in Zimmermann 1869]) – Leconte (1850, 1861, 1863, 1866, 1874, 1881); with one species, *Elmis concolor*, oddly described by Leconte (1881) in Le Conte & Horn (1881)! The author of Larainae is Leconte (1861) while the author of its type genus (*Lara*) is Le Conte (1852).

Similarly, two spelling versions exist for: Guérin-Méneville (1835) – Guérin Méneville (1843), and for Motschoulsky (1851, 1869) – Motschulsky (1853, 1854, 1860).

Macronychus caucasicus was described by Victor (1839). In fact, Victor is to be regarded as a pseudonym of Motschoulsky/Motschulsky (1810–1871), who occasionally used a pseudonym, because he had problems to get his manuscripts published (see http://species.wikimedia.org/wiki/Victor_Ivanovitsch_Motschulsky). However, without doubt, the correct authorship for *Macronychus caucasicus* is Victor, not Motschoulsky or Motschulsky. Unfortunately, in the "Catalogue of Palaearctic Coleoptera" (Jäch et al. 2006) the authorship of this species was changed to Motschulsky by the editors of the "Catalogue of Palaearctic Coleoptera" (Löbl & Smetana).

Castelnau (1840) described two species of Elmidae from Europe: *Elmis confusa* and *E. caliginosa*. In the "Catalogue of Palaearctic Coleoptera" (Jäch et al. 2006) the correct authority of these species was unfortunately changed to Laporte by the editors of the "Catalogue of Palaearctic Coleoptera" (Löbl & Smetana). In any case, the correct author name for *Elmis confusa* and *E. caliginosa* undoubtedly is Castelnau (= François-Louis Comte de Castelnau, 1802–1880). See Evenhuis (2012) for a detailed biography of Castelnau.

Publication Dates

Numerous publication dates, which were usually incorrectly cited in previous catalogues, are corrected herein.

Some examples: *Ampumixis* Sanderson, 1953a [not 1954], *Cleptelmis* Sanderson, 1953a [not 1954], *Esolus filum* (Fairmaire, 1871) [not 1870], *Gonielmis* Sanderson, 1953a [not 1954], *Hexanchorus dimorphus* Spangler & Staines, 2004b [not 2003], *Luchoelmis* Spangler & Staines, 2004a [not 2002], *Microdinodes jeanneli* Delève, 1946 [not 1945], *Promoresia* Sanderson, 1953a [not 1954], *Xenelmis tarsalis* Hinton, 1940a [not 1946]. In case of *Potamophilinus orientalis* (Guérin-Méneville, 1835), the year of description remains enigmatic and is based on a suggestion by Cowan (1971: 29).

Type Localities

Howard E. Hinton (1912–1977) described about 280 species and subspecies of Elmidae, mostly from South America. Unfortunately, he created some geographical confusion concerning species, which he collected in the Brazilian State of Rondônia, by actually attributing some of these locations (Guajará-Mirim, Porto Velho, Rio Candeias, Rio Ji-Paraná) to the State of Mato Grosso, especially in his earlier works (see Hinton 1940b, e, 1941a, 1945a–b, 1946a, 1971a, b, 1972d). However, according to Hinton's itinerary it is absolutely clear, that he collected these specimens in the State of Rondônia, not in Mato Grosso. For instance, Hinton (1945b: 92) described the type locality of *Elsianus amazonicus* [*Macrelmis*] as "Matto [sic] Grosso: Porto Velho, i.ix.1937 (H.E. Hinton)". Although there is at least one Porto Velho also in the State of Mato Grosso, it is clear from Hinton's itinerary, that on September 1, 1937 he was in the State of Rondônia. Many years later, in the description of *Neoelmis thyas* Hinton, 1972b, the type specimens of which were collected on the same day as those of *Elsianus amazonicus*, Hinton (1972b: 122) cited the type locality correctly: "Rondônia: Porto Velho, i.ix.1937 (H.E. Hinton)". Hinton probably never collected in the State of Mato Grosso, and all species, which he described as being from Mato Grosso, actually were collected in the State of Rondônia. Similarly, Hinton (1939e, 1971b) confused the Brazilian States of Amazonas and Pará. In total, 21 species are involved: *Austrolimnius pusio* Hinton, 1941a, *Cylloepus olcnus* Hinton, 1945a, *Gyrelmis longipes* Hinton, 1940b, *G. nubila* Hinton, 1940b, *G. pusio* Hinton, 1940b, *G. simplex* Hinton, 1940b, *G. thoracica basalis* Hinton, 1940b, *Hintonelmis atys* Hinton, 1971a, *H. carus* Hinton, 1971a, *H. opis* Hinton, 1971a, *H. sul* Hinton, 1971a, *Macrelmis amazonica* (Hinton, 1945b), *Neoelmis ampla* Hinton, 1940e, *N. marmorata* Hinton, 1940e, *Neolimnius palpalis* Hinton, 1939e, *Phanoceroides aquaticus* Hinton, 1939a, *Pilielmis abdera* Hinton, 1971b, *P. clita* Hinton, 1971b, *Tyletelmis mila* Hinton, 1972d, *Xenelmis tarsalis* Hinton, 1940a, *X. teres* Hinton, 1946a.

During the compilation of this catalogue we found that the type localities of two African species were hitherto assigned to wrong countries: the type locality of *Stenelmis aloysiisabaudiae* Pic, 1930 lies in Ethiopia [not Somalia], and the type locality of *Potamodytes antennatus* (Dohrn, 1882) lies in Ghana [not Guinea].

Availability and Authorship of Taxa Described by Yang & Zhang in Zhang & Yang (1995)

Zhang & Yang (1995) formally described five species of *Stenelmis* from China, one of which, *S. sinuata*, was a primary homonym, and was renamed *S. insufficiens* by Jäch & Kodada (2006). Apart from these five formal descriptions the publication by Zhang & Yang (1995) contains names of eight additional new species of *Stenelmis*, all ascribed to Yang & Zhang [in Zhang & Yang (1995)]. These names were mentioned in the key (*S. euronotana* [misspelled as "*euronotara*" in the English summary], *S. grossimarginata*, *S. indepresa* [incorrect original spelling of *indepressa*], *S. sinica*), or in the running text (*S. heteromorpha*, *S. huangkengana*, *S. montana*, *S. venticarinata*). Although they were not unambiguously indicated as intentionally new, diagnostic characters were provided and therefore these names must be regarded available, because they were published before the year 2000 (ICZN 1999: Art. 16). Formal descriptions of seven of these eight species were published later, by Zhang, Yang & Li (1997), Yang & Zhang (2002), and Zhang, Su & Yang (2003a) respectively; no formal description was ever published in case of *S. heteromorpha*. For *S. euronotana* even two formal descriptions were published: by Yang & Zhang (2002) and (under reversed authorship: "Zhang & Yang"!) by Zhang & Yang (2003). In the formal description of *S. grossimarginata* the authorship was also published in reversed order ("Zhang & Yang"), and in the formal description of *S. montana* the authorship was ascribed to Zhang, Su & Yang.

Internet Lists

There are various internet sources claiming to provide global information on taxa names: e.g.

"Thomson Reuters Index to Organism Names" (ION) (http://www.organismnames.com/)

"Global Names Index" (http://gni.globalnames.org/)

"Encyclopedia of Life" (EOL) (http://eol.org/pages/40962533/overview)

"Global Biodiversity Information Facility" (GBIF) (http://www.gbif.org/)

All these lists are grossly incomplete and their data are quite unreliable and often faulty.

We randomly selected two genera, *Ancyronyx* and *Neoelmis*, to test "Thomson Reuters Index to Organism Names" (ION) and the "Encyclopedia of Life" (EOL). In ION, 42 species of *Neoelmis* are listed, while 49 species and one subspecies do in fact exist; authorship is not mentioned in about one third of the names listed, and the names of at least three species are misspelled, making it in at least one case difficult to assign it to an existing species ("*N. upera*" [no author] might refer to *N. aspera* Hinton, 1940a). The species of *Ancyronyx* found in ION are a mess, with several species listed even three times, and several species are listed, which in fact belong to *Pseudancyronyx*. EOL is also far from reliable. *Neoelmis apicalis angusta* Hinton, 1939d and *N. argentinensis* Manzo & Archangelsky, 2012 are missing, and one species (*N. saon* Hinton, 1972c) is listed a second time, under the name "*N. soon*" (remarkably, the same misspelling is found also in ION). In the genus *Ancyronyx*, 16 species are listed in EOL, while 23 plus one subspecies are actually known.

Data provided by GBIF are similarly flawy, with remarkable parallels to the other lists.

The name of the type species of Elmidae (*Elmis maugetii*) is spelled incorrectly ("*maugetti*") in EOL and GBIF.

The elmid genus *Potamodytes* is erroneously listed under Dryopidae in ION, EOL, and in GBIF!

In several internet lists (e.g. EOL, GBIF, ITIS, Wikipedia) we detected nomina nuda erroneously treated as available names (e.g. *Stenelmis florala, S. harleyi, S. williami*).

Too many cooks spoil the broth, and it seems that all internet lists use the same cookbook, in which the most important pages are missing.

In ZooBank (http://zoobank.org/) only very (!) few species of Elmidae are registered at all. The year of publication for *Hydora musci* Lambert, Maier & Leschen, 2015 is incorrectly cited as 2014 (date of online publication) – see: http://zoobank.org/NomenclaturalActs/D2F3089F-CE00-4256-9D8F-021D02838062. The name *Hydora musci* is available only since the date of the publication of the print version (16.VII.2015), and therefore this species is not listed in this catalogue (deadline for extant taxa: 31.XII.2014).

Names published electronically

According to an amendment of the International Code of Zoological Nomenclature new names published online are available for the purpose of zoological nomenclature since 2012, provided they meet several conditions. For instance, they must be produced by a method that assures (!) widely accessible electronic copies with fixed content and layout (e.g. PDF/A). However, as it was impressively demonstrated by Dubois et al. (2013) PDF format does evidently not assure fixed content and layout.

Therefore we did not include *Hydora musci* Lambert, Maier & Leschen, 2015 in this catalogue (deadline for extant taxa: 31.XII.2014). This species was published online in 2014 (see: http://zoobank.org/NomenclaturalActs/D2F3089F-CE00-4256-9D8F-021D02838062) but the printed version was published on 16.VII.2015.

New Nomenclatorial and Taxonomic Acts

Authors of new acts are indicated in square brackets.

Change in Rank

Protelmidae Jeannel, 1950 (**new rank**) elevated to family rank from tribal rank (Protelmini)

[M.A. Jäch, J. Kodada & F. Čiampor]

The species of the four protelmid genera differ from true elmids in a number of characters (e.g. absence of frontoclypeal suture; presence of specialized sensilla on maxillary palpus; different structure of mesoventrite; elytra with unique apical interlocking tongue; lack of a typical elmid ovipositor) (Čiampor, Čiamporová-Zaťovičová, Jäch & Kodada, in prep.). Therefore we decided to treat them as a separate family, which is probably not the sister of Elmidae.

New Substitute Name

Helminthocharis filicornis Jäch & Kodada **nom.n.** for *Elmis nitidula* Fairmaire, 1897b [junior primary homonym of *Elmis nitidula* Leconte, 1866 (*Oulimnius*)]

[M.A. Jäch & J. Kodada]

New Synonymies

Cylloepus sexualis Hinton, 1937a **syn.n.** of *C. abnormis* (Horn, 1870)

[P.J. Spangler (†) & H.P. Brown (†)]

Synonymy based on an unpublished checklist compiled by P.J. Spangler (†) & H.P. Brown (†).

Elmis lousisi (Mařan, 1939) **syn.n.** of *E. bosnica* (Zaitzev, 1908)

[M.A. Jäch]

A total of 24 type specimens of *Elmis lousisi* (20 deposited in the Národní muzeum, Prague, Czech Republic, and four deposited in the Naturhistorisches Museum Wien, Austria) have been examined by M.A. Jäch in 2014.

Limnius latiusculus (Zaitzev, 1947) **syn.n.** of *L. opacus* Müller, 1806a

[M.A. Jäch]

Synonymy based on original description and on examination of material from Armenia.

Limnius rambouseki (Mařan, 1939) **syn.n.** of *L. perrisi* (Dufour, 1843)
[M.A. Jäch]
Two type specimens of *Limnius rambouseki*, deposited in the Národní muzeum, Prague, Czech Republic, have been examined by M.A. Jäch in 2014.

Normandia Pic, 1900a **syn.n.** of *Riolus* Mulsant & Rey, 1872
[M.A. Jäch]
Examination of *Riolus somcheticus* (Kolenati, 1846) revealed that in this species several generic characters (length of styli of ovipositor, structure of ventral sac of aedeagus) are intermediate between *Normandia* and *Riolus* and therefore would not allow its unambiguous generic assignation if both genera were retained as good genera.

Oulimnius lacustris (Stephens, 1828) **syn.n.** of *O. tuberculatus* (Müller, 1806a)
[M.A. Jäch]
Four type specimens (incl. a male) of *Oulimnius lacustris*, deposited in the Natural History Museum, London, U.K., have been examined by M.A. Jäch in 2014.

Riolus syriacus (Allard, 1869) **syn.n.** of *R. somcheticus* (Kolenati, 1846)
[M.A. Jäch]
A female syntype of *Riolus somcheticus*, labelled: "Tiflis", "Kolenati", deposited in the Naturhistorisches Museum Wien, Austria, was examined by M.A. Jäch in 2014.

New Combinations

Graphelmis nigromaculata (Chûjô & Satô, 1964) **comb.n.**, transferred from *Stenelmis* Dufour, 1835
[M.A. Jäch]
The original description of *Stenelmis nigromaculata* provides clear evidence that this species in fact belongs to *Graphelmis*.

Heterelmis gibbosa (Grouvelle, 1889b) **comb.n.**, transferred from *Elmis* Latreille, 1802
[M.A. Jäch]
One female syntype, deposited in the Muséum national d'Histoire naturelle, Paris, France, was examined by M.A. Jäch in 2014.

Hexacylloepus danforthi (Musgrave, 1935) **comb.n.**, *H. haitianus* (Darlington, 1936) **comb.n.**, and *H. lahottensis* (Darlington, 1936) **comb.n.**, transferred from *Cylloepus* Erichson, 1847

[W.D. Shepard]

Photographs of the holotypes were examined by W.D. Shepard.

Peloriolus brunneus (F.H. Waterhouse, 1879) **comb.n.**, transferred from *Elmis* Latreille, 1802

[M.A. Jäch]

Two syntypes, deposited in the Natural History Museum, London, U.K., have been examined by M.A. Jäch in 2013.

The species transferred herein from *Normandia* to *Riolus* are not regarded as new combinations, because they all had previously been used in combination with *Riolus*.

Designations of Type Species

Helminthocharis picea Grouvelle, 1906c, designated as type species of *Helminthocharis* Grouvelle, 1906c

[M.A. Jäch & J. Kodada]

Helminthopsis lucida Grouvelle, 1906c, designated as type species of *Helminthopsis* Grouvelle, 1906c

[M.A. Jäch & J. Kodada]

Mandatory Corrections of Incorrect Original Spellings of Valid Names (ICZN 1999: Art. 31.1.2, 32.5, 34.2)

Austrelmis patagonica Manzo & Archangelsky, 2012 (incorrect original spelling: "*patagonicus*")

[M.A. Jäch]

Hedyselmis gibbosa Jäch & Boukal, 1997b (incorrect original spelling: "*gibbosus*")

[M.A. Jäch]

Podelmis atra Jäch, 1982b (incorrect original spelling: "*ater*")
[M.A. Jäch]

Stenelmis gauglerae Bollow, 1941 (incorrect original spelling: "*gaugleri*")
[M.A. Jäch]

Stenelmis punctatissima Bollow, 1940b (incorrect original spelling: "*punctatissimus*")
[M.A. Jäch]

Stenelmis roiae Bollow, 1941 (incorrect original spelling: "*roii*")
[M.A. Jäch]

Stenelmis semifumosa Hinton, 1936e (incorrect original spelling: "*semifumosus*")
[M.A. Jäch]

Family Group Names

The phylogeny of Elmidae has never been revised with modern cladistic methods. The generally acknowledged concept of two subfamilies (Elminae and Larainae) has not been confirmed by DNA analyses. In fact, based on DNA sequencing, several recent papers suggest that the phylogenetic position of Larainae is inside Elminae (see e.g. Čiampor & Ribera 2006, Hayashi et al. 2013, Kundrata et al. 2013).

In the list below, the hierarchical positions of taxa below the family level (subfamilies, tribes, subtribes) follow recent concepts (e.g. Bouchard et al. 2011) and do not necessarily reflect opinions of the authors of this catalogue.

Elmidae Curtis, 1830: pl. 294. Type genus: *Elmis* Latreille, 1802. NOTE: Name conserved by International Commission on Zoological Nomenclature (ICZN 1995: Opinion 1812), following proposal by Jäch (1994b); invalid spellings: "Limniidae" (Stephens 1828: 104), "Helmidae" (Grouvelle 1900b: 268), "Helminthidae" (Ganglbauer 1904a: 108), "Elminthidae" (Steffan 1958: 127), "Elmididae" (Madge & Pope 1980: 257).

Elminae Curtis, 1830: pl. 294. NOTE: Incorrect subsequent spellings: "Helminide" (Bollow 1940a: 117), "Elmitae" (Delève 1946: 324).

Elmini Curtis, 1830: pl. 294.

Elmina Curtis, 1830: pl. 294.

Stenelmina Mulsant & Rey, 1872: 49. Type genus: *Stenelmis* Dufour, 1835.

Macronychini Gistel, 1848: [unnumbered page between columns 400 and 409]. Type genus: *Macronychus* Müller, 1806b. NOTE: Authorship erroneously ascribed to Mulsant & Rey (1872) by various authors.

Ancyronychini Ganglbauer, 1904a: 108. Type genus: *Ancyronyx* Erichson, 1847.

 | DOI 10.1163/9789004291775_002

Larainae Leconte, 1861: 116. Type genus: *Lara* Le Conte, 1852. NOTE: Originally named "Larinae", but, following proposals by Spangler (1986, 1987), the International Commission on Zoological Nomenclature (ICZN 1988: Opinion 1515) ruled the correct stem to be "Lara-" (instead of "Lar-") to avoid homonymy with Laridae Rafinesque, 1815: 72 (Type Genus: *Larus* Linnæus, 1758: 136) in Aves; incorrect subsequent spellings: "Lavinæ" (Hinton, 1937c: 289), "Laritae" (Delève 1946: 323).

Laraini Leconte, 1861: 116.

Potamophilini Mulsant & Rey, 1872: 11. Type genus *Potamophilus* Germar, 1811.

Protelmidae Jeannel, 1950: 170 (*new rank*). Type genus: *Protelmis* Grouvelle, 1911c. NOTE: This name was not listed in Bouchard et al. (2011); hitherto regarded as a tribe of Elminae.

Genus and Species Group Names

All genera, species and subspecies are listed in alphabetical order. Synonyms are listed chronologically.

A total of 147 extant genera of Elmidae is presently recognized (deadline: 31.XII.2014). Most of these genera were not erected on the basis of cladistic analyses and have hitherto not been confirmed by molecular studies. One Palearctic genus, *Normandia*, is synonymized herein, based on morphological characteristics. Numerous other genera are very similar to each other or hardly distinguishable morphologically; e.g. *Narpus* and *Neoriohelmis*; *Elmidolia* and *Ilamelmis*; *Heterelmis*, *Promoresia* and *Optioservus* (see Kamite 2012); *Notriolus* and *Simsonia* (see Glaister 1999); *Ordobrevia* and *Stenelmis* (see Jäch 1984d: 284–285, Hayashi et al. 2013). We may expect a larger number of new synonymies once these genera are studied with modern methods, using also molecular data. Several genera described from Madagascar (e.g. *Aspidelmis*, *Elmidolia*, *Exolimnius*, and *Pseudelmidolia*) seem to be identical. Most of the genera, which were based solely on the reduced eyes will probably be found to be junior synonyms; for instance, the author of the Haitian genera *Anommatelmis* and *Lemalelmis* himself stated that they "apparently are derived" from certain extant species of an existing genus (Spangler 1981a), which means that these genera should be synonymized.

While a number of genera certainly will have to be synonymized, some especially speciose genera, such as *Cylloepus* and *Stenelmis* may have to be split up into several ones.

Numerous genera are in need of revision (e.g. *Austrolimnius*, *Cylloepus*, *Elmis*, *Esolus*, *Limnius*, *Macrelmis*, *Microcylloepus*, *Neoelmis*, *Stenelmis*).

Only 14 subgenera are recognized in Elmidae at present. None of these were based on cladistic analyses and none have ever been confirmed by molecular studies. Correct subgeneric assignation of species has been found to be often impossible. Therefore all subgenera are listed alphabetically under the respective genus name followed by the alphabetical list of species.

Synonyms are listed in chronological order.

Initials of first names are given only for two authors, C. O. Waterhouse (1879) and F. H. Waterhouse (1879), because they share the same surname and the same year of publication.

Distributional data: Generally, only country names are given. For some of the larger countries (e.g. Australia, Brazil, Canada, India, USA) also provinces,

 | DOI 10.1163/9789004291775_003

states or territories are indicated, while for certain other countries (e.g. Japan, Philippines) data about islands (or island groups) are provided; for Indonesia data are provided about islands and/or provinces. The Autonomous Republic of Crimea is included in Ukraine.

Abbreviations:

Australia (ACT)	Australia (Australian Capital Territory)
Congo (DR)	Democratic Republic of the Congo [formerly: Zaire]
Congo (R)	Republic of the Congo
Korea (R)	South Korea (Republic of Korea)
Korea (DPR)	North Korea (Democratic People's Republic of Korea)

Some country names are followed by a reference between brackets, referring either to questionable records, to unpublished data (e.g. theses, or papers in preparation) or to little known or rarely cited publications, or recently published papers.

Some country names are placed between brackets; in these countries the respective species almost certainly do occur, but have not yet officially been recorded, and no specimens from there are known to us.

Elmidae

Extant taxa

Currently, 147 genera and 1498 species are recognized. Deadline: 31.XII.2014.

Aesobia Jäch, 1982b

Aesobia Jäch, 1982b: 97. — Type species: *Aesobia pygmaea* Jäch, 1982b. — Gender feminine.

pygmaea Jäch, 1982b

Aesobia pygmaea Jäch, 1982b: 97

TYPE LOCALITY: Sri Lanka.

DISTRIBUTION: Sri Lanka.

NOTE: In the original publication the epithet name was used four times: two times spelled as "*pygmea*" (p. 97) and two times spelled as "*pygmaea*" (pp. 110, 113); Jäch (1984d) used only the name "*pygmaea*" and can therefore be regarded as "First Reviser" (ICZN 1999: Art. 24.2.4), determining the correct original spelling to be "*pygmaea*"; the name "*pygmea*" in the original description is to be regarded as a lapsus calami.

Ampumixis Sanderson, 1953a

Ampumixis Sanderson, 1953a: 155. — Type species: *Helmis dispar* Fall, 1925. — Gender feminine.

NOTE: Originally described in a key, formal description published by Sanderson (1954: 3).

dispar (Fall, 1925)

Helmis dispar Fall, 1925: 180

TYPE LOCALITY: USA (California).

DISTRIBUTION: USA (California, Oregon, Washington).

Ancyronyx Erichson, 1847

Ancyronyx Erichson, 1847: 522. — Type species: *Macronychus variegatus* Germar, 1824. — Gender masculine (ICZN 1999: Art. 30.1.2), see also Jäch (1994a: 619).

acaroides acaroides Grouvelle, 1896d

Ancyronyx acaroides Grouvelle, 1896d: 50

TYPE LOCALITY: Indonesia (Sumatra).

DISTRIBUTION: Brunei, Indonesia (Java, Sumatra), Laos, Malaysia (Penang, Perak, Sabah, Sarawak), Myanmar, Vietnam.

acaroides cursor Jäch, 1994a

Ancyronyx acaroides cursor Jäch, 1994a: 607

Type Locality: Indonesia (Bali).

Distribution: Indonesia (Bali).

buhid Freitag, 2013

Ancyronyx buhid Freitag, 2013: 49

Type Locality: Philippines (Mindoro).

Distribution: Philippines (Mindoro).

helgeschneideri Freitag & Jäch, 2007

Ancyronyx helgeschneideri Freitag & Jäch, 2007: 55

Type Locality: Philippines (Palawan).

Distribution: Philippines (Busuanga, Palawan).

hjarnei Jäch, 2003

Ancyronyx hjarnei Jäch, 2003: 255

Type Locality: Indonesia (Sulawesi).

Distribution: Indonesia (Sulawesi).

jaechi Freitag, 2012

Ancyronyx jaechi Freitag, 2012: 60

Type Locality: Sri Lanka.

Distribution: Sri Lanka.

johanni Jäch, 1994a

Ancyronyx johanni Jäch, 1994a: 607

Type Locality: Indonesia (Siberut).

Distribution: Indonesia (Siberut).

malickyi Jäch, 1994a

Ancyronyx malickyi Jäch, 1994a: 609

Type Locality: Indonesia (Sumatra).

Distribution: Indonesia (Sumatra), Laos, Malaysia (Kelantan, Perak, Sabah, Sarawak), Thailand.

minerva Freitag & Jäch, 2007

Ancyronyx minerva Freitag & Jäch, 2007: 50

Type Locality: Philippines (Palawan).

Distribution: Philippines (Mindoro, Palawan); erroneously recorded from Busuanga by Freitag (2013: 43).

minutulus Freitag & Jäch, 2007

Ancyronyx minutulus Freitag & Jäch, 2007: 53

Type Locality: Philippines (Palawan).

Distribution: Philippines (Palawan).

montanus Freitag & Balke, 2011

Ancyronyx montanus Freitag & Balke, 2011: 67

Type Locality: Philippines (Palawan).

Distribution: Philippines (Palawan).

Note: In the original publication the heading of the description erroneously reads: "*Ancyronyx montanus* Freitag sp. n." (Freitag & Balke 2011: 67), but the correct authority undoubtedly is Freitag & Balke (see Freitag & Balke 2011: 50, 55, 68, 71, 79, Freitag 2012: 64); in Wikispecies the authorship is incorrectly cited as "*Ancyronyx montanus* Freitag in Freitag & Balke, 2011".

patrolus Freitag & Jäch, 2007

Ancyronyx patrolus Freitag & Jäch, 2007: 41

Type Locality: Philippines (Palawan).

Distribution: Philippines (Busuanga, Palawan).

procerus Jäch, 1994a

Ancyronyx procerus Jäch, 1994a: 611

Type Locality: Malaysia (Sarawak).

Distribution: Brunei, Malaysia (Pahang, Sarawak), Philippines (Busuanga), Vietnam.

pseudopatrolus Freitag & Jäch, 2007

Ancyronyx pseudopatrolus Freitag & Jäch, 2007: 46

Type Locality: Philippines (Palawan).

Distribution: Philippines (Palawan).

pulcherrimus Kodada, Jäch & Čiampor, 2014

Ancyronyx pulcherrimus Kodada, Jäch & Čiampor, 2014: 386

Type Locality: Brunei.

Distribution: Brunei.

punkti Freitag & Jäch, 2007

Ancyronyx punkti Freitag & Jäch, 2007: 47

Type Locality: Philippines (Palawan).

Distribution: Philippines (Palawan).

raffaelacatharina Jäch, 2004

Ancyronyx raffaelacatharina Jäch, 2004: 392

Type Locality: Indonesia (Sulawesi).

Distribution: Indonesia (Sulawesi).

reticulatus Kodada, Jäch & Čiampor, 2014

Ancyronyx reticulatus Kodada, Jäch & Čiampor, 2014: 384

Type Locality: Malaysia (Sabah).

Distribution: Malaysia (Sabah).

sarawacensis Jäch, 1994a

Ancyronyx sarawacensis Jäch, 1994a: 614

Type Locality: Malaysia (Sarawak).

Distribution: Malaysia (Sabah, Sarawak).

schillhammeri Jäch, 1994a

Ancyronyx schillhammeri Jäch, 1994a: 617

Type Locality: Philippines (Mindoro).

Distribution: Philippines (Mindoro).

sophiemarie Jäch, 2004

Ancyronyx sophiemarie Jäch, 2004: 389

Type Locality: Philippines (Sibuyan).

Distribution: Philippines (Sibuyan).

tamaraw Freitag, 2013

Ancyronyx tamaraw Freitag, 2013: 43

Type Locality: Philippines (Mindoro).

Distribution: Philippines (Bohol?, Luzon?, Mindoro).

variegatus (Germar, 1824)

Macronychus variegatus Germar, 1824: 89

Type Locality: North America (details unknown).

Distribution: Canada (Ontario, Quebec), USA (Alabama, Arkansas, Connecticut, Delaware, District of Columbia, Florida, Georgia, Illinois, Indiana, Kansas, Kentucky, Louisiana, Maine, Maryland, Michigan, Mississippi, Missouri, New Jersey, New York, North Carolina, North Dakota, Ohio, Oklahoma, Pennsylvania, Rhode Island, South Carolina, Tennessee, Texas, Vermont, Virginia, West Virginia, Wisconsin).

Note: Incorrect subsequent spelling: "*variegata*" (Brown 1972a: 14, 1983: 3).

Synonym: *Elmis cincta* Say, 1825: 186. — Type locality: USA (Pennsylvania). Note: Incorrect original spelling: "*cinctus*".

yunju Bian, Guo & Ji, 2012

Ancyronyx yunju Bian, Guo & Ji, 2012: 58

Type Locality: China (Jiangxi).

Distribution: China (Jiangxi), Laos, Vietnam.

Anommatelmis Spangler, 1981a

Anommatelmis Spangler, 1981a: 376. — Type species: *Anommatelmis botosaneanui* Spangler, 1981a. — Gender feminine.

Note: *Anommatelmis* is obviously a junior synonym of *Hexacylloepus* Hinton, 1940a (see Spangler 1981a: 379: "*Anommatelmis botosaneanui* and the endemic ... *Cylloepus haitianus* [*Hexacylloepus haitianus* (Darlington, 1936)] ... are similar in most character states and *A. botosaneanui* appears to be derived from that taxon").

botosaneanui Spangler, 1981a

Anommatelmis botosaneanui Spangler, 1981a: 377

Type Locality: Haiti.

Distribution: Haiti.

Aspidelmis Delève, 1954

Aspidelmis Delève, 1954: 29. — Type species: *Aspidelmis scutellaris* Delève, 1954. — Gender feminine.

grouvellei Delève, 1964a

Aspidelmis grouvellei Delève, 1964a: 33

Type Locality: Madagascar.

Distribution: Madagascar.

perrieri (Fairmaire, 1897a)

Elmis perrieri Fairmaire, 1897a: 98

Type Locality: Madagascar.

Distribution: Madagascar.

scutellaris Delève, 1954

Aspidelmis scutellaris Delève, 1954: 30

Type Locality: Madagascar.

Distribution: Madagascar.

subfuliginosa (Grouvelle, 1906a)

Elmis subfuliginosa Grouvelle, 1906a: 149

Type Locality: Madagascar.

Distribution: Madagascar.

Atractelmis Chandler, 1954

Atractelmis Chandler, 1954: 125. — Type species: *Atractelmis wawona* Chandler, 1954. — Gender feminine.

wawona Chandler, 1954

Atractelmis wawona Chandler, 1954: 125

Type Locality: USA (California).

Distribution: USA (California, Idaho, Oregon).

Aulacosolus Jäch & Boukal, 1997a

Aulacosolus Jäch & Boukal, 1997a: 207. — Type species: *Aulacosolus tenuior* Jäch & Boukal, 1997a. — Gender masculine.

bharatensis Jäch & Boukal, 1997a

Aulacosolus bharatensis Jäch & Boukal, 1997a: 213

Type Locality: India (Kerala).

Distribution: India (Kerala).

carinatus Jäch & Boukal, 1997a

Aulacosolus carinatus Jäch & Boukal, 1997a: 212

Type Locality: Thailand.

Distribution: Thailand.

scida Jäch & Boukal, 1997a

Aulacosolus scida Jäch & Boukal, 1997a: 212

Type Locality: Thailand.

Distribution: Thailand.

spinosus Jäch & Boukal, 1997a

Aulacosolus spinosus Jäch & Boukal, 1997a: 211

Type Locality: Malaysia (Kedah).

Distribution: Indonesia (Sumatra), Malaysia (Kedah), Thailand.

tarsalis Jäch & Boukal, 1997a

Aulacosolus tarsalis Jäch & Boukal, 1997a: 211

TYPE LOCALITY: Laos.

DISTRIBUTION: Laos.

tenuior Jäch & Boukal, 1997a

Aulacosolus tenuior Jäch & Boukal, 1997a: 210

TYPE LOCALITY: Thailand.

DISTRIBUTION: Laos, Thailand.

Austrelmis Brown, 1984

Austrelmis Brown, 1984: 126. — Type species: *Macrelmis leleupi* Delève, 1968a. — Gender feminine.

anthracina (Germain, 1892)

Elmis anthracina Germain, 1892: 245

TYPE LOCALITY: Chile.

DISTRIBUTION: Chile.

chilensis (Germain, 1854)

Elmis chilensis Germain, 1854: 327

TYPE LOCALITY: Chile.

DISTRIBUTION: Chile.

condimentarius (Philippi, 1864)

Elmis condimentarius Philippi, 1864: 96

TYPE LOCALITY: Peru.

DISTRIBUTION: Peru.

NOTE: Incorrect subsequent spelling: "*condimentaria*" (Hinton 1937b: 131); the epithet name can be regarded as an adjective (condimentarius, -a, -um [pertaining to a spice]) or as a noun in apposition (condimentarius, masculinum [spice merchant]), and therefore the original spelling ("*Elmis condimentarius*") is to be retained (ICZN 1999: Art. 31.2.2); no author is mentioned in the original description itself, but according to the table of contents of the journal volume (p. 441) the author (Philippi) is unambiguously indicated; according to Brown (1984: 128) this species is probably a senior synonym of one or more species described later from Peru; syntypes are deposited in the Chilean National Museum of Natural History, Santiago de Chile, but no type specimens of this species have been examined since the original description, however, in the original description (Philippi 1864: 93) the author stresses, that he has mounted (glued) several hundred specimens and provided almost all ("fast sämtliche") friends of the coleopterology with specimens; therefore, it can be assumed that at least some of the syntypes should still exist in various other historical museum collections; Rodolfo Amando Philippi (1808–1904) was curator at the Museo Nacional de Historia Natural de Chile, and a member of the Entomological Society of Stettin [Szczecin, Poland].

confluenta (Hinton, 1940i)

Macrelmis confluenta Hinton, 1940i: 126
TYPE LOCALITY: Peru.
DISTRIBUTION: Peru.

confusa (Hinton, 1940i)

Macrelmis confusa Hinton, 1940i: 125
TYPE LOCALITY: Peru.
DISTRIBUTION: Peru.

consors consors (Hinton, 1940i)

Macrelmis consors Hinton, 1940i: 131
TYPE LOCALITY: Peru.
DISTRIBUTION: Bolivia, Peru.

consors mooni (Hinton, 1940i)

Macrelmis consors mooni Hinton, 1940i: 134
TYPE LOCALITY: Bolivia.
DISTRIBUTION: Bolivia.
NOTE: It can be assumed from the original description, that this is rather a valid species or a synonym (hot spring variety) of *Macrelmis consors*, but not a subspecies.

costulata (Janssens, 1957)

Macrelmis costulata Janssens, 1957: 7
TYPE LOCALITY: Chile.
DISTRIBUTION: Chile.

dorotae Więźlak, 1987a

Austrelmis dorotae Więźlak, 1987a: 299
TYPE LOCALITY: Peru.
DISTRIBUTION: Peru.

elegans (Janssens, 1957)

Macrelmis elegans Janssens, 1957: 5
TYPE LOCALITY: Chile.
DISTRIBUTION: Chile.

flavitarsis (Grouvelle, 1896b)

Helmis flavitarsis Grouvelle, 1896b: 5 [(9)]

Type Locality: Uruguay.

Distribution: Bolivia (Więźlak 1987a), Uruguay.

gardineri (Hinton, 1940i)

Macrelmis gardineri Hinton, 1940i: 139

Type Locality: Bolivia.

Distribution: Bolivia.

gilsoni (Hinton, 1940i)

Macrelmis gilsoni Hinton, 1940i: 137

Type Locality: Peru.

Distribution: Bolivia, Peru.

glabra (Hinton, 1940i)

Macrelmis glabra Hinton, 1940i: 136

Type Locality: Peru.

Distribution: Peru.

lata (Hinton, 1940i)

Macrelmis lata Hinton, 1940i: 128

Type Locality: Peru.

Distribution: Peru.

leleupi (Delève, 1968a)

Macrelmis leleupi Delève, 1968a: 268

Type Locality: Ecuador.

Distribution: Ecuador.

patagonica Manzo & Archangelsky, 2012

Austrelmis patagonica Manzo & Archangelsky, 2012: 269

Type Locality: Argentina.

Distribution: Argentina.

Note: Incorrect original spelling: "*patagonicus*", herewith changed mandatorily (ICZN 1999: Art. 34.2).

peruana (Hinton, 1937b)

Macrelmis peruana Hinton, 1937b: 134

TYPE LOCALITY: Peru.

DISTRIBUTION: Bolivia, Peru.

steineri (Spangler, 1980b)

Macrelmis steineri Spangler, 1980b: 206

TYPE LOCALITY: Peru.

DISTRIBUTION: Peru.

thermarum (Hinton, 1940i)

Macrelmis thermarum Hinton, 1940i: 135

TYPE LOCALITY: Bolivia.

DISTRIBUTION: Bolivia.

tibialis (Grouvelle, 1896a)

Helmis tibialis Grouvelle, 1896a: 50

TYPE LOCALITY: Bolivia.

DISTRIBUTION: Bolivia, Chile.

woytkowskii (Hinton, 1937b)

Macrelmis woytkowskii Hinton, 1937b: 131

TYPE LOCALITY: Peru.

DISTRIBUTION: Bolivia, Peru.

SYNONYM: *Macrelmis woytkowskii bicolor* Janssens, 1957: 5. — Type locality: Bolivia. NOTE: Originally described as "*Macrelmis woytkowskii* var. n. *bicolor*"; might be a good species.

Austrolimnius Carter & Zeck, 1929

Austrolimnius Carter & Zeck, 1929: 61. — Type species: *Elmis polita* King, 1865. — Gender masculine.

NOTE: This genus strongly resembles *Helminthocharis* Grouvelle, 1906c.

SUBGENERA: ***Baltelmis*** Hinton, 1968: 99. — Type species: *Elmis punctulata* King, 1865. — Gender feminine.

Helonelmis Hinton, 1968: 99. — Type species: *Austrolimnius waterhousei* Hinton, 1965. — Gender feminine.

Helonoma Hinton, 1968: 100. — Type species: *Austrolimnius pusio* Hinton, 1941a. — Gender feminine.

Hintoniella Özdikmen, 2005: 233. — Type species: *Austrolimnius oblongus* Carter & Zeck, 1933. – Gender feminine. NOTE: Substitute name for *Helonastes* Hinton, 1968.

HOMONYM: *Helonastes* Hinton, 1968: 100. — Gender masculine? NOTE: Junior homonym of *Helonastes* Common, 1960: 327 (Lepidoptera).

Limnelmis Hinton, 1968: 102. — Type species: *Austrolimnius celsus* Hinton, 1965. — Gender feminine.

Neosolus Carter & Zeck, 1929: 68. — Type species: *Neosolus tropicus* Carter & Zeck, 1929. — Gender masculine.

Pelonaetes Hinton, 1968: 101. — Type species: *Austrolimnius atys* Hinton, 1965. — Gender masculine?

Telmatelmis Hinton, 1968: 102. — Type species: *Austrolimnius metasternalis* Carter & Zeck, 1938. — Gender feminine.

Tiphonaetes Hinton, 1968: 101. — Type species: *Austrolimnius hebrus* Hinton, 1965. — Gender masculine?

Tiphonelmis Hinton, 1968: 100. — Type species: *Helmis pilula* Grouvelle, 1889b. — Gender feminine.

alcine Hinton, 1965

Austrolimnius alcine Hinton, 1965: 154

TYPE LOCALITY: Australia (New South Wales).

DISTRIBUTION: Australia (ACT, New South Wales).

amanus Hinton, 1965

Austrolimnius amanus Hinton, 1965: 158

TYPE LOCALITY: Australia (New South Wales).

DISTRIBUTION: Australia (New South Wales).

NOTE: Epithet name probably is a noun in apposition.

anytus Hinton, 1965

Austrolimnius anytus Hinton, 1965: 169

TYPE LOCALITY: Australia (Victoria).

DISTRIBUTION: Australia (New South Wales, Victoria).

apicarinatus Boukal, 1997

Austrolimnius apicarinatus Boukal, 1997: 183

TYPE LOCALITY: Papua New Guinea.

DISTRIBUTION: Papua New Guinea.

araneus Jäch, 1985

Austrolimnius araneus Jäch, 1985: 250

TYPE LOCALITY: Papua New Guinea.

DISTRIBUTION: Papua New Guinea.

archidonanus Delève, 1968a

Austrolimnius archidonanus Delève, 1968a: 227

Type Locality: Ecuador.

Distribution: Ecuador.

Note: Incorrect subsequent spelling: "*archidonaus*" (Segura et al. 2013: 9).

ater (Grouvelle, 1889c)

Helmis atra Grouvelle, 1889c: 165

Type Locality: Venezuela.

Distribution: Venezuela.

atriceps Carter & Zeck, 1932

Austrolimnius atriceps Carter & Zeck, 1932: 203

Type Locality: Australia (Queensland).

Distribution: Australia (Queensland).

atys Hinton, 1965

Austrolimnius atys Hinton, 1965: 129

Type Locality: Australia (ACT).

Distribution: Australia (ACT).

balkei Boukal, 1997

Austrolimnius balkei Boukal, 1997: 168

Type Locality: Indonesia (Papua).

Distribution: Indonesia (Papua).

bispina Boukal, 1997

Austrolimnius bispina Boukal, 1997: 199

Type Locality: Indonesia (Papua).

Distribution: Indonesia (Papua), Papua New Guinea.

Note: Epithet name is a noun in apposition.

bocainensis Miranda, Sampaio & Passos, 2012

Austrolimnius bocainensis Miranda, Sampaio & Passos, 2012: 21

Type Locality: Brazil (São Paulo).

Distribution: Brazil (Rio de Janeiro, São Paulo).

brevior Boukal, 1997

Austrolimnius brevior Boukal, 1997: 190

Type Locality: Indonesia (Papua).

Distribution: Indonesia (Papua).

browni Hinton, 1971c

Austrolimnius browni Hinton, 1971c: 98

Type Locality: Venezuela.

Distribution: Venezuela.

brunneus Jäch, 1985

Austrolimnius brunneus Jäch, 1985: 249

Type Locality: Papua New Guinea.

Distribution: Papua New Guinea.

bustardi Hinton, 1965

Austrolimnius bustardi Hinton, 1965: 138

Type Locality: Australia (ACT).

Distribution: Australia (ACT).

caicus Hinton, 1965

Austrolimnius caicus Hinton, 1965: 148

Type Locality: Australia (New South Wales).

Distribution: Australia (New South Wales).

capys Hinton, 1965

Austrolimnius capys Hinton, 1965: 144

Type Locality: Australia (New South Wales).

Distribution: Australia (New South Wales).

carus Hinton, 1965

Austrolimnius carus Hinton, 1965: 122

Type Locality: Australia (New South Wales).

Distribution: Australia (New South Wales).

celsus Hinton, 1965

Austrolimnius celsus Hinton, 1965: 136

Type Locality: Australia (ACT).

Distribution: Australia (ACT, New South Wales).

ceto Hinton, 1965

Austrolimnius ceto Hinton, 1965: 121

TYPE LOCALITY: Australia (New South Wales).

DISTRIBUTION: Australia (New South Wales).

cheops Hinton, 1965

Austrolimnius cheops Hinton, 1965: 120

TYPE LOCALITY: Australia (Victoria).

DISTRIBUTION: Australia (Victoria).

chiloensis (Champion, 1918)

Elmis chiloensis Champion, 1918: 48

TYPE LOCALITY: Chile.

DISTRIBUTION: Chile.

SYNONYM: *Elmis prothoracica* Hinton, 1936f: 56. — Type locality: Chile. NOTE: Proposed as (unjustified) substitute name for *Elmis chiloensis* Champion, 1918; *Elmis chiloensis* is not a homonym of *Elmis chilensis* Germain, 1854 (the epithet name *chiloensis* refers to the island of Chiloe).

codrus Hinton, 1965

Austrolimnius codrus Hinton, 1965: 118

TYPE LOCALITY: Australia (New South Wales).

DISTRIBUTION: Australia (New South Wales).

curtulus (Sharp, 1882)

Elmis curtula Sharp, 1882: 139

TYPE LOCALITY: Panama.

DISTRIBUTION: Panama.

NOTE: Incorrect original spelling: "*Elmis curtulus*".

dayi Hinton, 1965

Austrolimnius dayi Hinton, 1965: 161

TYPE LOCALITY: Australia (ACT).

DISTRIBUTION: Australia (ACT, Victoria).

dentibialis Boukal, 1997

Austrolimnius dentibialis Boukal, 1997: 169

TYPE LOCALITY: Indonesia (Papua).

DISTRIBUTION: Indonesia (Papua), Papua New Guinea.

didas Hinton, 1965

Austrolimnius didas Hinton, 1965: 143

TYPE LOCALITY: Australia (New South Wales).

DISTRIBUTION: Australia (ACT, New South Wales, Victoria).

diemenensis Carter & Zeck, 1935

Austrolimnius diemenensis Carter & Zeck, 1935: 79

TYPE LOCALITY: Australia (Tasmania).

DISTRIBUTION: Australia (ACT, New South Wales, Tasmania, Victoria).

difficilis Boukal, 1997

Austrolimnius difficilis Boukal, 1997: 188

TYPE LOCALITY: Indonesia (Papua).

DISTRIBUTION: Indonesia (Papua).

dudgeoni Boukal, 1997

Austrolimnius dudgeoni Boukal, 1997: 171

TYPE LOCALITY: Papua New Guinea.

DISTRIBUTION: Papua New Guinea.

egregius Delève, 1968a

Austrolimnius egregius Delève, 1968a: 230

TYPE LOCALITY: Ecuador.

DISTRIBUTION: Ecuador, Venezuela.

ekari Boukal, 1997

Austrolimnius ekari Boukal, 1997: 195

TYPE LOCALITY: Indonesia (Papua).

DISTRIBUTION: Indonesia (Papua).

elatus Hinton, 1941a

Austrolimnius elatus Hinton, 1941a: 161

TYPE LOCALITY: Chile.

DISTRIBUTION: Chile.

eris Hinton, 1971c

Austrolimnius eris Hinton, 1971c: 94

TYPE LOCALITY: Mexico.

DISTRIBUTION: Argentina, Belize, Brazil (Goiás) Costa Rica, Ecuador, Guatemala, Mexico, Panama, Venezuela.

evansi **Hinton, 1965**

Austrolimnius evansi Hinton, 1965: 152

Type Locality: Australia (New South Wales).

Distribution: Australia (New South Wales).

exilis **Jäch, 1985**

Austrolimnius exilis Jäch, 1985: 247

Type Locality: Papua New Guinea.

Distribution: Papua New Guinea.

fallax **Hinton, 1965**

Austrolimnius fallax Hinton, 1965: 123

Type Locality: Australia (New South Wales).

Distribution: Australia (New South Wales).

festus **Hinton, 1965**

Austrolimnius mormo festus Hinton, 1965: 159

Type Locality: Australia (New South Wales).

Distribution: Australia (New South Wales).

formosus **(Sharp, 1882)**

Elmis formosa Sharp, 1882: 140

Type Locality: Guatemala.

Distribution: Argentina, Belize, Brazil (Rio de Janeiro), Costa Rica, Guatemala, Mexico, Nicaragua, Panama, Peru, Venezuela.

Note: Incorrect original spelling: "*Elmis formosus*".

foveatus **Boukal, 1997**

Austrolimnius foveatus Boukal, 1997: 172

Type Locality: Papua New Guinea.

Distribution: Papua New Guinea.

gurakor **Boukal, 1997**

Austrolimnius gurakor Boukal, 1997: 173

Type Locality: Papua New Guinea.

Distribution: Papua New Guinea.

halffteri **Hinton, 1972a**

Austrolimnius halffteri Hinton, 1972a: 137

Type Locality: Mexico.

Distribution: Mexico.

hebrus Hinton, 1965

Austrolimnius hebrus Hinton, 1965: 162

TYPE LOCALITY: Australia (Victoria).

DISTRIBUTION: Australia (Victoria).

hera Hinton, 1965

Austrolimnius hera Hinton, 1965: 129

TYPE LOCALITY: Australia (New South Wales).

DISTRIBUTION: Australia (New South Wales).

hercules Boukal, 1997

Austrolimnius hercules Boukal, 1997: 196

TYPE LOCALITY: Papua New Guinea.

DISTRIBUTION: Papua New Guinea.

hilum Boukal, 1997

Austrolimnius hilum Boukal, 1997: 193

TYPE LOCALITY: Papua New Guinea.

DISTRIBUTION: Papua New Guinea.

insinuatus Boukal, 1997

Austrolimnius insinuatus Boukal, 1997: 191

TYPE LOCALITY: Indonesia (Papua).

DISTRIBUTION: Indonesia (Papua).

isdellensis Zeck, 1948

Austrolimnius isdellensis Zeck, 1948: 277

TYPE LOCALITY: Australia (Western Australia).

DISTRIBUTION: Australia (Western Australia).

isus Hinton, 1965

Austrolimnius isus Hinton, 1965: 170

TYPE LOCALITY: Australia (New South Wales).

DISTRIBUTION: Australia (New South Wales, Victoria).

jaechi Boukal, 1997

Austrolimnius jaechi Boukal, 1997: 201

TYPE LOCALITY: Indonesia (Ceram).

DISTRIBUTION: Indonesia (Ambon, Ceram).

keyi Hinton, 1965

Austrolimnius keyi Hinton, 1965: 160

Type Locality: Australia (New South Wales).

Distribution: Australia (New South Wales).

lacrimabilis Boukal, 1997

Austrolimnius lacrimabilis Boukal, 1997: 192

Type Locality: Papua New Guinea.

Distribution: Papua New Guinea.

laevigatus (Grouvelle, 1889b)

Helmis laevigata Grouvelle, 1889b: 407

Type Locality: Brazil (Rio de Janeiro).

Distribution: Brazil (Rio de Janeiro, Santa Catarina, São Paulo).

Note: Incorrect original spelling: "*lævigata*".

luridus Carter & Zeck, 1929

Austrolimnius luridus Carter & Zeck, 1929: 62

Type Locality: Australia (New South Wales).

Distribution: Australia (New South Wales, Victoria).

mackerrasi Hinton, 1965

Austrolimnius mackerrasi Hinton, 1965: 152

Type Locality: Australia (New South Wales).

Distribution: Australia (New South Wales).

maro Hinton, 1965

Austrolimnius maro Hinton, 1965: 141

Type Locality: Australia (New South Wales).

Distribution: Australia (New South Wales, Victoria).

medon Hinton, 1965

Austrolimnius medon Hinton, 1965: 139

Type Locality: Australia (Victoria).

Distribution: Australia (Victoria).

menopon Hinton, 1965

Austrolimnius menopon Hinton, 1965: 169

Type Locality: Australia (New South Wales).

Distribution: Australia (New South Wales).

messa Hinton, 1965

Austrolimnius messa Hinton, 1965: 142

Type Locality: Australia (ACT).

Distribution: Australia (ACT).

metasternalis Carter & Zeck, 1938

Austrolimnius metasternalis Carter & Zeck, 1938: 170

Type Locality: Australia (Victoria).

Distribution: Australia (New South Wales, Victoria).

mila Hinton, 1965

Austrolimnius mila Hinton, 1965: 141

Type Locality: Australia (ACT).

Distribution: Australia (ACT).

miletus Hinton, 1965

Austrolimnius miletus Hinton, 1965: 139

Type Locality: Australia (New South Wales).

Distribution: Australia (New South Wales).

minutus Delève, 1968a

Austrolimnius minutus Delève, 1968a: 225

Type Locality: Ecuador.

Distribution: Ecuador.

montanus (King, 1865)

Elmis montana King, 1865: 160

Type Locality: Australia (New South Wales).

Distribution: Australia (ACT, New South Wales, Queensland).

Note: Incorrect original spelling: "*Elmis montanus*".

mormo Hinton, 1965

Austrolimnius mormo Hinton, 1965: 158

Type Locality: Australia (ACT).

Distribution: Australia (ACT, New South Wales, Victoria).

mucubajiensis Gómez & Bello, 2006

Austrolimnius mucubajiensis Gómez & Bello, 2006: 14

Type Locality: Venezuela.

Distribution: Venezuela.

musgravei Hinton, 1939b

Austrolimnius musgravei Hinton, 1939b: 197

TYPE LOCALITY: French Guiana.

DISTRIBUTION: French Guiana.

nicon Hinton, 1965

Austrolimnius nicon Hinton, 1965: 135

TYPE LOCALITY: Australia (New South Wales).

DISTRIBUTION: Australia (New South Wales).

nomia Hinton, 1965

Austrolimnius nomia Hinton, 1965: 132

TYPE LOCALITY: Australia (New South Wales).

DISTRIBUTION: Australia (ACT, New South Wales).

nyctelioides (Germain, 1892)

Elmis nyctelioides Germain, 1892: 247

TYPE LOCALITY: Chile.

DISTRIBUTION: Argentina, Chile.

oblongus Carter & Zeck, 1933

Austrolimnius oblongus Carter & Zeck, 1933: 372

TYPE LOCALITY: Australia (New South Wales).

DISTRIBUTION: Australia (New South Wales).

ochus Hinton, 1965

Austrolimnius ochus Hinton, 1965: 144

TYPE LOCALITY: Australia (New South Wales).

DISTRIBUTION: Australia (New South Wales).

olenus Hinton, 1965

Austrolimnius olenus Hinton, 1965: 151

TYPE LOCALITY: Australia (New South Wales).

DISTRIBUTION: Australia (ACT, New South Wales).

opis Hinton, 1965

Austrolimnius opis Hinton, 1965: 126

TYPE LOCALITY: Australia (New South Wales).

DISTRIBUTION: Australia (ACT, New South Wales).

papuanus Jäch, 1985

Austrolimnius papuanus Jäch, 1985: 252

Type Locality: Papua New Guinea.

Distribution: Papua New Guinea.

pictus Jäch, 1985

Austrolimnius pictus Jäch, 1985: 247

Type Locality: Papua New Guinea.

Distribution: Papua New Guinea.

pilulus (Grouvelle, 1889b)

Helmis pilula Grouvelle, 1889b: 406

Type Locality: Brazil (Santa Catarina).

Distribution: Brazil (Rio de Janeiro, Santa Catarina, São Paulo).

politus (King, 1865)

Elmis polita King, 1865: 160

Type Locality: Australia (New South Wales).

Distribution: Australia (New South Wales).

Note: Incorrect original spelling: "*Elmis politus*".

punctulatus aeolus Hinton, 1965

Austrolimnius punctulatus aeolus Hinton, 1965: 116

Type Locality: Australia (New South Wales).

Distribution: Australia (New South Wales).

punctulatus punctulatus (King, 1865)

Elmis punctulata King, 1865: 161

Type Locality: Australia (New South Wales).

Distribution: Australia (New South Wales).

Note: Incorrect original spelling: "*Elmis punctulatus*".

pusio Hinton, 1941a

Austrolimnius pusio Hinton, 1941a: 157

Type Locality: Brazil (Rondônia); erroneously, Hinton (1941a: 159) attributed the type locality to the State of Mato Grosso.

Distribution: Brazil (Rondônia), Ecuador.

ramuensis Boukal, 1997

Austrolimnius ramuensis Boukal, 1997: 161

Type Locality: Papua New Guinea.

Distribution: Papua New Guinea.

resa Hinton, 1965

Austrolimnius resa Hinton, 1965: 125

Type Locality: Australia (New South Wales).

Distribution: Australia (New South Wales, Victoria).

schoelleri Jäch, 1982b

Austrolimnius schoelleri Jäch, 1982b: 89

Type Locality: Indonesia (Papua).

Distribution: Indonesia (Papua).

schoelleroides Jäch, 1985

Austrolimnius schoelleroides Jäch, 1985: 251

Type Locality: Papua New Guinea.

Distribution: Papua New Guinea.

seductor Boukal, 1997

Austrolimnius seductor Boukal, 1997: 175

Type Locality: Papua New Guinea.

Distribution: Papua New Guinea.

similis Boukal, 1997

Austrolimnius similis Boukal, 1997: 182

Type Locality: Papua New Guinea.

Distribution: Papua New Guinea.

solitarius Boukal, 1997

Austrolimnius solitarius Boukal, 1997: 176

Type Locality: Papua New Guinea.

Distribution: Papua New Guinea.

speculifer Boukal, 1997

Austrolimnius speculifer Boukal, 1997: 177

Type Locality: Indonesia (Papua).

Distribution: Indonesia (Papua).

suffusus Carter & Zeck, 1930

Austrolimnius luridus suffusus Carter & Zeck, 1930: 190

Type Locality: Australia (New South Wales).

Distribution: Australia (New South Wales).

Note: Originally described as a variety of *Austrolimnius luridus* ("*Austrolimnius luridus*, C. & Z., var. *suffusus*, n. var.") by Carter & Zeck (1930: 190); formal description published by Carter & Zeck (1935: 80) under "*Austrolimnius suffusus* n.sp.".

sul Hinton, 1965

Austrolimnius sul Hinton, 1965: 133

Type Locality: Australia (New South Wales).

Distribution: Australia (New South Wales).

sulcicollis (Sharp, 1882)

Elmis sulcicollis Sharp, 1882: 139

Type Locality: Panama.

Distribution: Costa Rica, Guatemala, Mexico, Panama.

sulmo Hinton, 1965

Austrolimnius sulmo Hinton, 1965: 159

Type Locality: Australia (New South Wales).

Distribution: Australia (ACT, New South Wales, Victoria).

tarsalis Hinton, 1941a

Austrolimnius tarsalis Hinton, 1941a: 159

Type Locality: Bolivia.

Distribution: Bolivia, Ecuador.

thyas Hinton, 1965

Austrolimnius thyas Hinton, 1965: 126

Type Locality: Australia (New South Wales).

Distribution: Australia (New South Wales).

troilus Hinton, 1965

Austrolimnius troilus Hinton, 1965: 166

Type Locality: Australia (ACT).

Distribution: Australia (ACT, New South Wales).

tropicus (Carter & Zeck, 1929)

Neosolus tropicus Carter & Zeck, 1929: 69

Type Locality: Australia (Northern Territory).

DISTRIBUTION: Australia (Northern Territory).

SYNONYM: *Neosolus tropicus asper* Carter & Zeck, 1929: 69. — Type locality: Australia (Northern Territory). NOTE: Originally described as "var. *asper*" under *Neosolus tropicus*; might be a good species.

ullrichi Jäch, 1985

Austrolimnius ullrichi Jäch, 1985: 246

TYPE LOCALITY: Papua New Guinea.

DISTRIBUTION: Indonesia (Papua), Papua New Guinea.

uncatus Miranda, Sampaio & Passos, 2012

Austrolimnius uncatus Miranda, Sampaio & Passos, 2012: 15

TYPE LOCALITY: Brazil (São Paulo).

DISTRIBUTION: Brazil (Rio de Janeiro, São Paulo).

variabilis Carter & Zeck, 1932

Austrolimnius variabilis Carter & Zeck, 1932: 203

TYPE LOCALITY: Australia (Queensland).

DISTRIBUTION: Australia (ACT, New South Wales, Queensland).

victoriensis Carter & Zeck, 1929

Austrolimnius victoriensis Carter & Zeck, 1929: 61

TYPE LOCALITY: Australia (Victoria).

DISTRIBUTION: Australia (ACT, New South Wales, Victoria).

virilis Boukal, 1997

Austrolimnius virilis Boukal, 1997: 186

TYPE LOCALITY: Papua New Guinea.

DISTRIBUTION: Papua New Guinea.

waterhousei Hinton, 1965

Austrolimnius waterhousei Hinton, 1965: 127

TYPE LOCALITY: Australia (New South Wales).

DISTRIBUTION: Australia (ACT, New South Wales).

weylandensis Boukal, 1997

Austrolimnius weylandensis Boukal, 1997: 178

TYPE LOCALITY: Indonesia (Papua).

DISTRIBUTION: Indonesia (Papua).

wigglesworthi Hinton, 1965

Austrolimnius wigglesworthi Hinton, 1965: 128

Type Locality: Australia (New South Wales).

Distribution: Australia (New South Wales).

Bryelmis Barr, 2011

Bryelmis Barr, 2011: 198. — Type species: *Bryelmis rivularis* Barr, 2011. — Gender feminine.

idahoensis Barr, 2011

Bryelmis idahoensis Barr, 2011: 200

Type Locality: USA (Idaho).

Distribution: USA (Idaho).

rivularis Barr, 2011

Bryelmis rivularis Barr, 2011: 204

Type Locality: USA (Washington).

Distribution: USA (Oregon, Washington).

siskiyou Barr, 2011

Bryelmis siskiyou Barr, 2011: 207

Type Locality: USA (California).

Distribution: USA (California, Oregon).

Cephalolimnius Delève, 1973a

Cephalolimnius Delève, 1973a: 18. — Type species: *Cephalolimnius ater* Delève, 1973a. — Gender masculine.

ater Delève, 1973a

Cephalolimnius ater Delève, 1973a: 19

Type Locality: Sri Lanka.

Distribution: Sri Lanka.

Cleptelmis Sanderson, 1953a

Cleptelmis Sanderson, 1953a: 155. — Type species: *Helmis ornata* Schaeffer, 1911 (= *Cleptelmis addenda* Fall, 1907). — Gender feminine.

Note: Originally described in a key; formal description published by Sanderson (1954: 4).

addenda (Fall, 1907)

Elmis addenda Fall, 1907: 226

TYPE LOCALITY: USA (New Mexico).

DISTRIBUTION: Canada (Alberta, British Columbia), USA (Arizona, California, Colorado, Idaho, Montana, Nevada, New Mexico, Oregon, South Dakota, Utah, Washington, Wyoming).

NOTE: Incorrect original spelling: "*Elmis addendus*".

SYNONYM: *Helmis ornata* Schaeffer, 1911: 120. — Type locality: USA (Montana). NOTE: Incorrect original spelling: "*ornatus*"; synonymized by Shepard (1998: 291).

Coxelmis Carter & Zeck, 1929

Coxelmis Carter & Zeck, 1929: 67. — Type species: *Elmis novemnotata* King, 1865. — Gender feminine.

novemnotata (King, 1865)

Elmis novemnotata King, 1865: 159

TYPE LOCALITY: Australia (New South Wales).

DISTRIBUTION: Australia (New South Wales).

NOTE: Incorrect original spelling: "*Elmis novem-notatus*".

trinotata Carter & Zeck, 1929

Coxelmis trinotata Carter & Zeck, 1929: 67

TYPE LOCALITY: Australia (New South Wales).

DISTRIBUTION: Australia (New South Wales).

v-fasciata (Lea, 1895)

Elmis v-fasciata Lea, 1895: 590

TYPE LOCALITY: Australia (New South Wales).

DISTRIBUTION: Australia (New South Wales).

NOTE: Incorrect subsequent spelling: "*v. fasciata*" (Carter & Zeck 1929: 69, 70).

Ctenelmis Delève, 1964b

Ctenelmis Delève, 1964b: 171. — Type species: *Ctenelmis harrisoni* Delève, 1964b. — Gender feminine.

SUBGENUS: ***Paractenelmis*** Delève, 1964d: 532. — Type species: *Ctenelmis discrepans* Delève, 1964d. — Gender feminine.

crinipes Delève, 1966b

Ctenelmis crinipes Delève, 1966b: 99

Type Locality: Angola.

Distribution: Angola.

discrepans **Delève, 1964d**

Ctenelmis discrepans Delève, 1964d: 534 (subgenus *Paractenelmis*)

Type Locality: South Africa.

Distribution: South Africa.

elegans **Delève, 1966b**

Ctenelmis elegans Delève, 1966b: 104

Type Locality: South Africa.

Distribution: Angola, South Africa.

harrisoni **Delève, 1964b**

Ctenelmis harrisoni Delève, 1964b: 171

Type Locality: South Africa.

Distribution: Angola, South Africa.

incerta **(Grouvelle, 1890)**

Helmis incerta Grouvelle, 1890: CCXII[212]

Type Locality: South Africa.

Distribution: Angola, South Africa.

lata **Delève, 1964d**

Ctenelmis lata Delève, 1964d: 532

Type Locality: South Africa.

Distribution: South Africa.

rufipes **Delève, 1966b**

Ctenelmis rufipes Delève, 1966b: 100

Type Locality: South Africa.

Distribution: Angola, South Africa.

tibialis **Delève, 1966b**

Ctenelmis tibialis Delève, 1966b: 102

Type Locality: South Africa.

Distribution: Angola, South Africa.

Cuspidevia Jäch & Boukal, 1995a

Cuspidevia Jäch & Boukal, 1995a: 301. — Type species: *Cuspidevia velaris* Jäch & Boukal, 1995a. — Gender feminine.

brevis Bian & Ji, 2010

Cuspidevia brevis Bian & Ji, 2010: 54

TYPE LOCALITY: China (Jiangxi).

DISTRIBUTION: China (Jiangxi).

jaechi Bian & Ji, 2010

Cuspidevia jaechi Bian & Ji, 2010: 56

TYPE LOCALITY: China (Jiangxi).

DISTRIBUTION: China (Jiangxi).

velaris Jäch & Boukal, 1995a

Cuspidevia velaris Jäch & Boukal, 1995a: 302

TYPE LOCALITY: China (Hong Kong).

DISTRIBUTION: China (Hong Kong).

Cylloepus Erichson, 1847

Cylloepus Erichson, 1847: 521. — Type species: *Limnius araneolus* Müller, 1806a. — Gender masculine.

NOTE: Incorrect subsequent spellings: "*Cyllœpus*" (Grouvelle 1889b: 394–399), "*Cyllœpus*" (Hinton 1945a: 43–67, 1946c: 713–733), "*Cylhepus*" (http://www.organismnames.com; http://gni.globalnames.org).

abnormis (Horn, 1870)

Elmis abnormis Horn, 1870: 38

TYPE LOCALITY: USA (Arizona).

DISTRIBUTION: Mexico, USA (Arizona, Texas).

SYNONYM: *Cylloepus sexualis* Hinton, 1937a: 102 **syn.n.** — Type locality: Mexico.

alcine Hinton, 1945a

Cylloepus alcine Hinton, 1945a: 62 ("*Cyllœpus*")

TYPE LOCALITY: Brazil (Santa Catarina).

DISTRIBUTION: Brazil (Santa Catarina).

araneolus (Müller, 1806a)

Limnius araneolus Müller, 1806a: 202

Type Locality: Peru.

Distribution: Peru.

Note: Incorrect subsequent spelling: "*areolus*" (Lacordaire 1854: 510).

atys Hinton, 1946c

Cylloepus atys Hinton, 1946c: 718 ("*Cyllœpus*")

Type Locality: Peru.

Distribution: Peru.

barberi Hinton, 1934

Cylloepus barberi Hinton, 1934: 193

Type Locality: Guatemala.

Distribution: Belize, Costa Rica, Guatemala.

bartolozzii Monte & Mascagni, 2012

Cylloepus bartolozzii Monte & Mascagni, 2012: 4

Type Locality: Ecuador.

Distribution: Ecuador.

blairi Hinton, 1936a

Cylloepus blairi Hinton, 1936a: 1

Type Locality: Mexico.

Distribution: Mexico.

brasiliensis Grouvelle, 1889b

Cylloepus brasiliensis Grouvelle, 1889b: 398 ("*Cyllœpus*")

Type Locality: Brazil (Santa Catarina).

Distribution: Brazil (Santa Catarina).

caicus Hinton, 1946c

Cylloepus caicus Hinton, 1946c: 723 ("*Cyllœpus*")

Type Locality: Peru.

Distribution: Peru.

Note: Incorrect subsequent spelling: "*caïcus*" (Janssens 1956: 2).

carinulus Hinton, 1945a

Cylloepus carinulus Hinton, 1945a: 53 ("*Cyllœpus*")

Type Locality: Brazil (Santa Catarina).

Distribution: Brazil (Santa Catarina).

cesari Monte & Mascagni, 2012

Cylloepus cesari Monte & Mascagni, 2012: 7

Type Locality: Ecuador.

Distribution: Ecuador.

confusus Hinton, 1936f

Cylloepus confusus Hinton, 1936f: 54

Type Locality: Brazil (Santa Catarina).

Distribution: Brazil (Santa Catarina).

Note: Substitute name for *Helmis concolor* Grouvelle, 1889b.

Homonyms: *Helmis concolor* Grouvelle, 1889b: 401. Note: Formerly a junior secondary homonym of *Elmis concolor* Leconte, 1881 (*Narpus*); permanently invalid (ICZN 1999: Art. 59.3).

Helmis sharpi Zaitzev, 1910: 28. Note: Substitute name for *Helmis concolor* Grouvelle, 1889b; junior secondary homonym of *Cylloepus sharpi* Grouvelle, 1889b; permanently invalid (ICZN 1999: Art. 59.3).

consobrinus Grouvelle, 1896a

Cylloepus consobrinus Grouvelle, 1896a: 42

Type Locality: Bolivia.

Distribution: Bolivia.

didas Hinton, 1945a

Cylloepus didas Hinton, 1945a: 59 ("*Cyllœpus*")

Type Locality: Brazil (Santa Catarina).

Distribution: Brazil (Santa Catarina).

dorvillei Passos & Felix, 2004b

Cylloepus dorvillei Passos & Felix, 2004b: 181

Type Locality: Brazil (Rio de Janeiro).

Distribution: Brazil (Rio de Janeiro).

drymus Hinton, 1946c

Cylloepus drymus Hinton, 1946c: 715 ("*Cyllœpus*")

Type Locality: Peru.

Distribution: Peru.

fabianorum Monte & Mascagni, 2012

Cylloepus fabianorum Monte & Mascagni, 2012: 10

Type Locality: Ecuador.

Distribution: Ecuador.

francescae Monte & Mascagni, 2012

Cylloepus francescae Monte & Mascagni, 2012: 11

Type Locality: Ecuador.

Distribution: Ecuador.

friburguensis Sampaio, Passos & Ferreira, 2011

Cylloepus friburguensis Sampaio, Passos & Ferreira, 2011: 61

Type Locality: Brazil (Rio de Janeiro).

Distribution: Brazil (Rio de Janeiro).

gigas Grouvelle, 1889b

Cylloepus gigas Grouvelle, 1889b: 397 ("*Cyllœpus*")

Type Locality: Brazil (Santa Catarina).

Distribution: Brazil (Santa Catarina).

gnidus Hinton, 1946c

Cylloepus gnidus Hinton, 1946c: 727 ("*Cyllœpus*")

Type Locality: Peru.

Distribution: Peru.

gounellei (Grouvelle, 1889b)

Helmis gounellei Grouvelle, 1889b: 400

Type Locality: Brazil (Santa Catarina).

Distribution: Brazil (Santa Catarina).

Note: According to Hinton (1936f: 55) this species "should be placed ... with the group of *C. ferruginea* [sic] (Horn)", which is currently a member of the genus *Hexacylloepus* Hinton, 1940a.

hastatus Delève, 1968a

Cylloepus hastatus Delève, 1968a: 258

Type Locality: Ecuador.

Distribution: Ecuador.

heteroceros (Sharp, 1882)

Elmis heteroceros Sharp, 1882: 135

Type Locality: Guatemala.

Distribution: Belize, Costa Rica, Guatemala, Mexico.

Note: Epithet name is a noun in apposition; incorrect subsequent spellings: "*heterocera*" (Zaitzev 1908: 302, 1910: 25), "*heterocerus*" (Hinton 1940a: 352).

macrelmoides Delève, 1968a

Cylloepus macrelmoides Delève, 1968a: 266

Type Locality: Ecuador.

Distribution: Ecuador, Venezuela.

maro Hinton, 1945a

Cylloepus maro Hinton, 1945a: 63 ("*Cyllœpus*")

Type Locality: Brazil (Santa Catarina).

Distribution: Brazil (Santa Catarina).

mazzai Monte & Mascagni, 2012

Cylloepus mazzai Monte & Mascagni, 2012: 13

Type Locality: Ecuador.

Distribution: Ecuador.

nelo Hinton, 1945a

Cylloepus nelo Hinton, 1945a: 58 ("*Cyllœpus*")

Type Locality: Brazil (São Paulo).

Distribution: Brazil (São Paulo); erroneously recorded from Santa Catarina by Segura et al. (2013: 14).

nessimiani Sampaio, Passos & Ferreira, 2011

Cylloepus nessimiani Sampaio, Passos & Ferreira, 2011: 60

Type Locality: Brazil (Rio de Janeiro).

Distribution: Brazil (Rio de Janeiro).

nicon Hinton, 1945a

Cylloepus nicon Hinton, 1945a: 55 ("*Cyllœpus*")

Type Locality: Brazil (Santa Catarina).

Distribution: Brazil (Santa Catarina).

olenus Hinton, 1945a

Cylloepus olenus Hinton, 1945a: 48 ("*Cyllœpus*")

Type Locality: Brazil (Amazonas).

Distribution: Brazil (Amazonas, Pará, Rondônia), French Guiana, Peru; erroneously recorded from the Brazilian State of Mato Grosso by Hinton (1945a: 50) and Segura et al. (2013: 14).

optatus Sharp, 1882

Cylloepus optatus Sharp, 1882: 129

Type Locality: Guatemala.

Distribution: Belize, Costa Rica, Guatemala, Mexico, Nicaragua, Panama.

palpalis palpalis Hinton, 1937b

Cylloepus palpalis Hinton, 1937b: 135

Type Locality: Peru.

Distribution: Peru.

palpalis tros Hinton, 1946c

Cylloepus palpalis tros Hinton, 1946c: 723 ("*Cyllœpus*")

Type Locality: Peru.

Distribution: Peru.

parallelus Delève, 1968a

Cylloepus parallelus Delève, 1968a: 263

Type Locality: Ecuador.

Distribution: Ecuador.

parkeri Sanderson, 1953b

Cylloepus parkeri Sanderson, 1953b: 38

Type Locality: USA (Arizona).

Distribution: USA (Arizona).

proximus Hinton, 1937a

Cylloepus proximus Hinton, 1937a: 106

Type Locality: Mexico.

Distribution: Mexico.

punctatus Hinton, 1940c

Cylloepus punctatus Hinton, 1940c: 395

Type Locality: Bolivia.

Distribution: Bolivia.

puncticollis (Hinton, 1934)

Stenelmis puncticollis Hinton, 1934: 198

Type Locality: Mexico.

Distribution: Mexico.

quinquecarinatus Sampaio, Passos & Ferreira, 2011

Cylloepus quinquecarinatus Sampaio, Passos & Ferreira, 2011: 58

Type Locality: Brazil (Rio de Janeiro).

Distribution: Brazil (Rio de Janeiro).

reitteri Grouvelle, 1889b

Cylloepus reitteri Grouvelle, 1889b: 399 ("*Cyllœpus*")

Type Locality: Brazil (Santa Catarina).

Distribution: Brazil (Rio de Janeiro, Santa Catarina, São Paulo).

sculpticollis Delève, 1968a

Cylloepus sculpticollis Delève, 1968a: 261

Type Locality: Ecuador.

Distribution: Colombia, Ecuador.

sharpi Grouvelle, 1889b

Cylloepus sharpi Grouvelle, 1889b: 398 ("*Cyllœpus*")

Type Locality: Brazil (Santa Catarina).

Distribution: Brazil (Santa Catarina).

silius Hinton, 1946c

Cylloepus silius Hinton, 1946c: 729 ("*Cyllœpus*")

Type Locality: Peru.

Distribution: Peru.

sparsus Hinton, 1940c

Cylloepus sparsus Hinton, 1940c: 403

Type Locality: Bolivia.

Distribution: Bolivia.

terzanii Monte & Mascagni, 2012

Cylloepus terzanii Monte & Mascagni, 2012: 16

Type Locality: Ecuador.

Distribution: Ecuador.

tuberculatus Hinton, 1940c

Cylloepus tuberculatus Hinton, 1940c: 400

Type Locality: Bolivia.

Distribution: Bolivia, Peru.

typhon Hinton, 1945a: 51

Cylloepus typhon Hinton, 1945a: 51 ("*Cyllœpus*")

TYPE LOCALITY: Brazil (Rio de Janeiro?); according to the original description the holotype "was probably collected in Rio de Janeiro" (Hinton 1945a: 52).

DISTRIBUTION: Brazil (Rio de Janeiro?).

ulpianus Hinton, 1946c

Cylloepus ulpianus Hinton, 1946c: 720 ("*Cyllœpus*")

TYPE LOCALITY: Peru.

DISTRIBUTION: Peru.

ventralis Hinton, 1940c

Cylloepus ventralis Hinton, 1940c: 397

TYPE LOCALITY: Bolivia.

DISTRIBUTION: Bolivia.

vianai Hinton, 1951

Cylloepus vianai Hinton, 1951: 820

TYPE LOCALITY: Argentina.

DISTRIBUTION: Argentina.

vicinus Hinton, 1940c

Cylloepus vicinus Hinton, 1940c: 407

TYPE LOCALITY: Bolivia.

DISTRIBUTION: Bolivia, Ecuador.

whitmanae Monte & Mascagni, 2012

Cylloepus whitmanae Monte & Mascagni, 2012: 18

TYPE LOCALITY: Ecuador.

DISTRIBUTION: Ecuador.

zagreus Hinton, 1945a

Cylloepus zagreus Hinton, 1945a: 64 ("*Cyllœpus*")

TYPE LOCALITY: Brazil (Minas Gerais).

DISTRIBUTION: Brazil (Minas Gerais).

Disersus Sharp, 1882

Disersus Sharp, 1882: 127. — Type species: *Disersus longipennis* Sharp, 1882. — Gender masculine.

ambocheilus Spangler & Santiago, 1987

Disersus ambocheilus Spangler & Santiago, 1987: 10

Type Locality: Ecuador.

Distribution: Ecuador, Peru.

cacicus (Coquerel, 1851)

Potamophilus cacicus Coquerel, 1851: 596

Type Locality: Colombia.

Distribution: Colombia, Ecuador.

chibcha Spangler & Santiago, 1987

Disersus chibcha Spangler & Santiago, 1987: 18

Type Locality: Colombia.

Distribution: Colombia, Venezuela.

dasycolus Spangler & Santiago, 1987

Disersus dasycolus Spangler & Santiago, 1987: 7

Type Locality: Ecuador.

Distribution: Ecuador, Venezuela.

inca Spangler & Santiago, 1987

Disersus inca Spangler & Santiago, 1987: 14

Type Locality: Ecuador.

Distribution: Colombia, Ecuador, Peru, Venezuela.

longipennis Sharp, 1882

Disersus longipennis Sharp, 1882: 127

Type Locality: Panama.

Distribution: Costa Rica, Panama.

pilitibia Spangler & Santiago, 1987

Disersus pilitibia Spangler & Santiago, 1987: 12

Type Locality: Colombia.

Distribution: Colombia, Ecuador.

quincemil Spangler & Santiago, 1987

Disersus quincemil Spangler & Santiago, 1987: 23

Type Locality: Peru.

Distribution: Peru.

saxicola Spangler & Santiago, 1987

Disersus saxicola Spangler & Santiago, 1987: 25

Type Locality: Ecuador.

Distribution: Ecuador.

uncus Spangler & Santiago, 1982

Disersus uncus Spangler & Santiago, 1982: 17

Type Locality: Costa Rica.

Distribution: Costa Rica, Panama.

Dryopomorphus Hinton, 1936d

Dryopomorphus Hinton, 1936d: 165. — Type species: *Dryopomorphus extraneus* Hinton, 1936d. — Gender masculine.

amami Yoshitomi & Satô, 2005

Dryopomorphus amami Yoshitomi & Satô, 2005: 466

Type Locality: Japan (Ryukyu Islands).

Distribution: Japan (Ryukyu Islands).

bishopi Hinton, 1971e

Dryopomorphus bishopi Hinton, 1971e: 295

Type Locality: Malaysia (Selangor).

Distribution: Malaysia (Kelantan, Pahang, Selangor).

extraneus Hinton, 1936d

Dryopomorphus extraneus Hinton, 1936d: 166

Type Locality: "Japan".

Distribution: Japan (Honshu, Kyushu, Shikoku).

grandis Čiampor, Čiamporová-Zaťovičová & Kodada, 2012

Dryopomorphus grandis Čiampor, Čiamporová-Zaťovičová & Kodada, 2012: 6

Type Locality: Malaysia (Sabah).

Distribution: Malaysia (Sabah).

hendrichi Čiampor & Kodada, 2006

Dryopomorphus hendrichi Čiampor & Kodada, 2006: 71

Type Locality: Malaysia (Johor).

Distribution: Malaysia (Johor).

jaechi Čiampor, Čiamporová-Zaťovičová & Kodada, 2012

Dryopomorphus jaechi Čiampor, Čiamporová-Zaťovičová & Kodada, 2012: 9

Type Locality: Malaysia (Sarawak).

Distribution: Malaysia (Sarawak).

laosensis Yoshitomi & Jeng, 2013

Dryopomorphus laosensis Yoshitomi & Jeng, 2013: 46

Type Locality: Laos.

Distribution: Laos.

memei Čiampor, Čiamporová-Zaťovičová & Kodada, 2012

Dryopomorphus memei Čiampor, Čiamporová-Zaťovičová & Kodada, 2012: 13

Type Locality: Malaysia (Sarawak).

Distribution: Brunei, Malaysia (Sarawak).

Note: The epithet name was partly misspelled ("*mamei*") in the original description (p. 14).

nakanei Nomura, 1958b

Dryopomorphus nakanei Nomura, 1958b: 45

Type Locality: Japan (Honshu).

Distribution: Japan (Honshu, Shikoku).

pekariki Čiampor, Čiamporová-Zaťovičová & Kodada, 2012

Dryopomorphus pekariki Čiampor, Čiamporová-Zaťovičová & Kodada, 2012: 4

Type Locality: Malaysia (Sabah).

Distribution: Malaysia (Sabah).

sarawacensis Čiampor, Čiamporová-Zaťovičová & Kodada, 2012

Dryopomorphus sarawacensis Čiampor, Čiamporová-Zaťovičová & Kodada, 2012: 12

Type Locality: Malaysia (Sarawak).

Distribution: Malaysia (Sarawak).

satoi Spangler, 1985b

Dryopomorphus satoi Spangler, 1985b: 416

Type Locality: Malaysia (Sabah).

Distribution: Brunei, Malaysia (Sabah, Sarawak).

siamensis Kodada, 1993a

Dryopomorphus siamensis Kodada, 1993a: 52

Type Locality: Thailand.

Distribution: Thailand.

yaku Yoshitomi & Satô, 2005

Dryopomorphus yaku Yoshitomi & Satô, 2005: 470

Type Locality: Japan (Yakushima).

Distribution: Japan (Tanegashima, Yakushima).

Dubiraphia Sanderson, 1954

Dubiraphia Sanderson, 1954: 4. — Type species: *Elmis quadrinotata* Say, 1825. — Gender feminine.

Note: The name was partly misspelled ("*Dubiraphai*") in the original description (p. 4); originally published in a key (Sanderson 1953a: 155), but unavailable, because no type species was designated (ICZN 1999: Art. 13.3).

bivittata (Le Conte, 1852)

Elmis bivittata Le Conte, 1852: 44

Type Locality: USA ("Upper Mississippi River").

Distribution: Canada (Alberta, Manitoba, Ontario, Quebec), USA (Illinois, Indiana, Wisconsin).

Note: Incorrect original spelling: "*Elmis bivittatus*".

brevipennis Hilsenhoff, 1973

Dubiraphia brevipennis Hilsenhoff, 1973: 60

Type Locality: USA (Louisiana).

Distribution: USA (Louisiana).

browni Hilsenhoff, 1973

Dubiraphia browni Hilsenhoff, 1973: 60

Type Locality: USA (Oklahoma).

Distribution: USA (Oklahoma).

brunnescens (Fall, 1925)

Helmis brunnescens Fall, 1925: 177

Type Locality: USA (California).

Distribution: USA (California).

Note: Incorrect subsequent spelling: "*brunnescans*" (Sanderson 1954: 5).

giulianii (Van Dyke, 1949)

Simsonia giulianii Van Dyke, 1949: 54

Type Locality: USA (California).

Distribution: USA (California).

Note: This species might be a junior synonym of *Dubiraphia brunnescens* (Shepard 1993: 3).

harleyi Barr, 1984

Dubiraphia harleyi Barr, 1984: 336

Type Locality: USA (Louisiana).

Distribution: USA (Louisiana).

minima Hilsenhoff, 1973

Dubiraphia minima Hilsenhoff, 1973: 59

Type Locality: USA (Wisconsin).

Distribution: Canada (Manitoba, Ontario, Quebec), USA (Indiana, Missouri, Ohio, Oklahoma, Wisconsin).

parva Hilsenhoff, 1973

Dubiraphia parva Hilsenhoff, 1973: 61

Type Locality: USA (Oklahoma).

Distribution: USA (Oklahoma).

quadrinotata (Say, 1825)

Elmis quadrinotata Say, 1825: 187

Type Locality: USA (details unknown, types probably lost).

Distribution: Canada (Manitoba, New Brunswick, Nova Scotia, Ontario, Prince Edward Island, Quebec), USA (Alabama, Illinois, Indiana, Maryland, Pennsylvania, Ohio, Oklahoma, Wisconsin).

Note: Incorrect original spelling: "*Elmis 4-notatus*".

robusta Hilsenhoff, 1973

Dubiraphia robusta Hilsenhoff, 1973: 59

Type Locality: USA (Wisconsin).

Distribution: USA (Wisconsin).

vittata (Melsheimer, 1844)

Elmis vittata Melsheimer, 1844: 99

Type Locality: USA (Pennsylvania).

Distribution: Canada (Manitoba, Ontario, Quebec), USA (Alabama, Arkansas, Delaware, District of Columbia, Florida, Georgia, Illinois, Indiana, Iowa, Kansas, Kentucky, Louisiana, Maryland, Minnesota, Mississippi, Missouri, Nebraska, New Jersey, New York, North Carolina, Ohio, Oklahoma, Pennsylvania, South Carolina, Tennessee, Texas, Virginia, West Virginia, Wisconsin).

Note: Incorrect original spelling: "*Elmis vittatus*".

Dupophilus Mulsant & Rey, 1872

Dupophilus Mulsant & Rey, 1872: 41. — Type species: *Dupophilus brevis* Mulsant & Rey, 1872. — Gender masculine.

brevis Mulsant & Rey, 1872

Dupophilus brevis Mulsant & Rey, 1872: 42

TYPE LOCALITY: France.

DISTRIBUTION: Armenia, France, Georgia, Italy, Portugal, Spain, Switzerland? (Schöll 2002), Turkey; the record from Switzerland is most doubtful (M. Hess, pers. comm.); the record from Germany (Przewoźny et al. 2011: 382) is most probably based on an error.

SYNONYMS: *Limnius gigas* Sharp, 1872: 263. — Type locality: Spain.

Latelmis insignis Reitter, 1885: 364. — Type locality: Georgia.

Elachistelmis Maier, 2012

Elachistelmis Maier, 2012: 62. — Type species: *Elachistelmis tetramera* Maier, 2012. — Gender feminine.

NOTE: Very close to *Neoelmis* Musgrave, 1935.

sipaliwiniensis Maier, 2012

Elachistelmis sipaliwiniensis Maier, 2012: 66

TYPE LOCALITY: Suriname.

DISTRIBUTION: French Guiana, Suriname.

tetramera Maier, 2012

Elachistelmis tetramera Maier, 2012: 63

TYPE LOCALITY: Suriname.

DISTRIBUTION: French Guiana, Suriname.

Elmidolia Fairmaire, 1897b

Elmidolia Fairmaire, 1897b: 369. — Type species: *Elmidolia sericans* Fairmaire, 1897b. — Gender feminine.

NOTE: Incorrect subsequent spellings: "*Helminthodolia*" Zaitzev (1908: 310) (not to be regarded as an emendation, see ICZN 1999: Art. 33.2.1), "*Elmidola*" (Fairmaire 1902: 343, 344).

SYNONYM: *Helminthopsoides* Delève, 1954: 33. — Type species: *Helmis binervosa* Grouvelle, 1899. — Gender feminine (ICZN 1999: Art. 30.1.4.4). NOTE: Incorrect original spelling: "*Helminthopsoïdes*"; synonymized by Delève (1963d: 2).

binervosa binervosa (Grouvelle, 1899)

Helmis binervosa Grouvelle, 1899: 184

TYPE LOCALITY: Madagascar.

DISTRIBUTION: Madagascar.

NOTE: Incorrect original spelling: "*binerrosa*", changed to *binervosa* by Zaitzev (1908: 301, 1910: 24); in the original description the epithet name was used only once (spelling: "*binerrosa*"), in the annual index of the volume (p. 836) the name is also given as "*binerrosa*", however, according to ICZN (1999: Art. 32.5.1) there is clear evidence of a printer's error, because the author described this species as being "bigranoso-carinatus" (diagnosis in Latin) and the elytra having two "carènes latérales" (description in French), clearly expressing the author's intention to name this species *binervosa*, a name being now in use for more than a hundred years.

binervosa lamarcquei (Paulian, 1959)

Helminthopsoides binervosa lamarcquei Paulian, 1959: 7

TYPE LOCALITY: Madagascar.

DISTRIBUTION: Madagascar.

fuliginea (Fairmaire, 1902)

Elmis fuliginea Fairmaire, 1902: 343

TYPE LOCALITY: Madagascar.

DISTRIBUTION: Madagascar.

NOTE: Incorrect original spelling: "*Elmis fuligineus*"; according to Delève (1964a: 30–35) it might be a senior synonym of *Elmidolia opaca* or *Aspidelmis grouvellei*; syntypes are deposited in the Muséum national d'Histoire naturelle, Paris, France, and in the Naturhistorisches Museum Wien, Austria.

opaca Delève, 1963d

Elmidolia opaca Delève, 1963d: 5

TYPE LOCALITY: Madagascar.

DISTRIBUTION: Madagascar.

sericans Fairmaire, 1897b

Elmidolia sericans Fairmaire, 1897b: 369

TYPE LOCALITY: Madagascar.

DISTRIBUTION: Madagascar.

Elmis Latreille, 1802

Elmis Latreille, 1802: 398. — Type species: *Elmis maugetii* Latreille, 1802. — Gender feminine.

NOTE: Stem (Elm-) and gender (feminine) conserved by ICZN (1995: Opinion 1812).

SYNONYMS: *Philydrus* Duftschmid, 1805: 304. — Type species: *Philydrus megerlei* Duftschmid, 1805 (*Elmis maugetii* Latreille, 1802). — Gender masculine. NOTE: Incorrect subsequent spellings: "*Pylidrus*" (Bertolini 1874: 102), "*Philhydrus*" (Brown 1981a: 145), the author of *Philydrus* was also spelled incorrectly ("Duftschmidt") in the same paper (Brown 1981a: 145).

Elminus Rafinesque, 1815: 112. — Type species: *Elmis maugetii* Latreille, 1802. — Gender masculine. NOTE: Unjustified emendation of *Elmis* Latreille, 1802 (see Rafinesque 1815: 223).

Lareynia Jaquelin du Val, 1859: 276. — Type species: *Elmis maugetii* Latreille, 1802: — Gender feminine. NOTE: Incorrect subsequent spellings: "*Larcynia*" (Kuwert 1889: 26), "*Lareyniea*" (Reiche 1879: 238, Sainte-Claire Deville 1905: 238).

Helmis Bedel, 1878: LXXV[75]. — Type species: *Elmis maugetii* Latreille, 1802: — Gender feminine. NOTE: Unjustified emendation of *Elmis* Latreille, 1802.

aenea (Müller, 1806a)

Limnius aeneus Müller, 1806a: 202

TYPE LOCALITY: Germany.

DISTRIBUTION: Albania, Andorra, Austria, Belarus, Belgium, Bosnia and Herzegovina, Bulgaria, Croatia (Mičetić Stanković et al. 2015: 98), Czech Republic, Denmark, Estonia, Finland, France, Germany, Hungary, Ireland, Italy, Latvia, Liechtenstein, Lithuania, Luxembourg, Montenegro (V. Mičetić Stanković, pers. comm.), Netherlands, Norway, Poland, Portugal, Romania, Russia, Slovakia, Slovenia, Spain, Sweden, Switzerland, Ukraine, United Kingdom.

NOTE: Incorrect subsequent spellings: "*Elmis æneus*" (Dejean 1821: 49, 1833: 131, Grenier 1863: 34), "*Elmis aeneus*" (Erichson 1847: 525, Gemminger 1851: 10, Bertolini 1874: 102).

SYNONYMS: *Elmis aenea affinis* Rey, 1889: 67. — Type locality: France. NOTE: Originally described as a variety of *Elmis aenea*.

Lareynia aenea alpina Kuwert, 1890: 30. — Type locality: "Alps". NOTE: Originally described as a variety of *Lareynia aenea*.

atlantis (Alluaud, 1922)

Helmis velutina atlantis Alluaud, 1922: 34

TYPE LOCALITY: Morocco.

DISTRIBUTION: Morocco.

bosnica assyriae Olmi, 1981

Elmis bosnica assyriae Olmi, 1981: 338

TYPE LOCALITY: Turkey.

DISTRIBUTION: Turkey.

NOTE: Status unclear.

bosnica bosnica (Zaitzev, 1908)

Helmis maugetii bosnica Zaitzev, 1908: 303

Type Locality: Bosnia and Herzegovina.

Distribution: Albania, Armenia, Bosnia and Herzegovina, Bulgaria, Croatia, Greece, Iran, Macedonia, Montenegro (V. Mičetić Stanković, pers. comm.), Russia, Serbia, Turkey.

Note: Substitute name for *Lareynia longicollis* Kuwert, 1890.

Homonym: *Lareynia longicollis* Kuwert, 1890: 48. Note: Junior secondary homonym of *Elmis longicollis* Sharp, 1882 (*Onychelmis*); permanently invalid (ICZN 1999: Art. 59.3).

Synonym: *Helmis lousisi* Mařan, 1939: 41 **syn.n.** — Type locality: Greece.

bosnica tenuis Berthélemy, 1979

Elmis bosnica tenuis Berthélemy, 1979: 25

Type Locality: Turkey.

Distribution: Greece, Turkey.

Note: Status unclear.

latreillei (Bedel, 1878)

Helmis latreillei Bedel, 1878: LXXV[75]

Type Locality: Europe, a lectotype has not been designated yet.

Distribution: Austria, Bosnia and Herzegovina, Bulgaria, Croatia, Czech Republic, France, Germany, Hungary, Italy, Luxembourg, Poland, Portugal, Romania, Slovakia, Slovenia, Spain, Switzerland, Ukraine.

Synonym: *Lareynia interrupta* Kuwert, 1890: 48. — Type locality: Spain.

maugetii fossulata (Kuwert, 1890)

Lareynia aenea fossulata Kuwert, 1890: 30

Type Locality: France (Corsica).

Distribution: France (Corsica), Italy (Sardinia).

Note: Originally described as a variety of *Lareynia aenea*; probably a discrete species.

maugetii maugetii Latreille, 1802

Elmis maugetii Latreille, 1802: 400

Type Locality: France.

Distribution: Armenia, Austria, Belarus, Belgium, Bosnia and Herzegovina, Bulgaria, Croatia, Czech Republic, Denmark, Estonia, France, Germany, Greece, Hungary, Italy, Latvia, Lithuania, Luxembourg, Macedonia, (Montenegro), Netherlands, Poland, Portugal, Romania, Russia, Serbia, Slovakia, Slovenia, Spain, Switzerland, Turkey, Ukraine.

Note: Incorrect subsequent spellings: "*maugeti*" (Bedel 1878: LXXV[75], Flach 1882: 252, Reitter 1883: 75, Zaitzev 1908: 302, 1910: 25), "*maugetti*" (e.g. Pakulnicka & Biesiadka 2011: 307, 308, 312); this species was named after the French zoologist René Maugé de Cely, who collected the type specimens (Maugé died on 21 February 1802 during an expedition to the Pacific led by Nicolas Baudin; he was taken ill at Timor and died when the expedition arrived in Tasmania, where he was buried on the point named after him by Baudin: Point Maugé on

Maria Island, off Tasmania's east coast); Latreille (1802) obviously latinized Maugé's name into Maugetius (genitiv: Maugetii) and therefore *maugetii* is the correct original spelling.

SYNONYMS: *Philydrus megerlei* Duftschmid, 1805: 305. — Type locality: Austria. NOTE: Regarded as a good species by Knie (1975a), as a subspecies of *Elmis maugetii* by Knie (1975b) and as variety of *Elmis maugetii* by Knie (1977, 1978); incorrect subsequent spelling: "*magelei*" (Tamutis et al. 2011: 215).

Elmis confusa Castelnau, 1840: 43. — Type locality: France. NOTE: Incorrect original spelling: "*Elmis confusus*".

Elmis kirschii Gerhardt, 1869: 261. — Type locality: Poland. NOTE: Incorrect subsequent spelling: "*kirschi*" (Zaitzev 1908: 303, 1910: 27, Mascagni & Calamandrei 1992: 130).

Helmis maugei Bedel, 1878: LXXV[75]. — Type locality: France. NOTE: Unjustified emendation of *Elmis maugetii* Latreille, 1802 (see above).

Lareynia similis Flach, 1882: 253. — Type locality: Spain.

Elmis media Rey, 1889: 67. — Type locality: France?

maugetii velutina Reiche, 1879

Elmis velutina Reiche, 1879: 238

TYPE LOCALITY: Algeria.

DISTRIBUTION: Algeria.

NOTE: Incorrect original spelling: "*velutinus*".

obscura (Müller, 1806a)

Limnius obscurus Müller, 1806a: 204

TYPE LOCALITY: Germany.

DISTRIBUTION: Austria, Bosnia and Herzegovina, Bulgaria, Croatia, Czech Republic, France, Germany, Greece, Hungary, Italy, Latvia, Macedonia, Netherlands, Poland, Russia, Slovakia, Slovenia, Switzerland.

NOTE: Incorrect subsequent spelling: "*Elmis obscurus*" (Dejean 1833: 131, Grenier 1863: 34).

SYNONYMS: *Elmis caliginosa* Castelnau, 1840: 43. — Type locality: France. NOTE: Incorrect original spelling: "*Elmis caliginosus*".

Lareynia croatica Kuwert, 1890: 49. — Type locality: Croatia, by lectotype ("holotype") designation (Berthélemy 1979: 30).

perezi Heyden, 1870

Elmis perezi Heyden, 1870: 110

TYPE LOCALITY: Spain.

DISTRIBUTION: France, Portugal, Spain.

quadricollis (Reitter, 1887)

Helmis quadricollis Reitter, 1887: 258

TYPE LOCALITY: Uzbekistan.

DISTRIBUTION: Afghanistan, Kazakhstan, Kyrgystan, Turkmenistan, Uzbekistan.

SYNONYM: *Elmis lindbergi* Janssens, 1959a: 7. — Type locality: Afghanistan.

rietscheli Steffan, 1958

Elmis rietscheli Steffan, 1958: 130

TYPE LOCALITY: Germany.

DISTRIBUTION: Austria, France, Germany, Hungary, Italy, Slovenia, Switzerland.

NOTE: Incorrect subsequent spelling: "*reitscheli*" (Nikitsky et al. 2010: 128).

rioloides (Kuwert, 1890)

Lareynia rioloides Kuwert, 1890: 49

TYPE LOCALITY: Bosnia and Herzegovina.

DISTRIBUTION: Albania, Austria, Belgium, Bosnia and Herzegovina, Bulgaria, Croatia (Mičetić Stanković et al. 2015: 100), Czech Republic, France, Germany, Greece, Hungary, Israel, Italy, Lebanon, Luxembourg, (Montenegro), Netherlands, Portugal, Romania, Russia, (Serbia), Slovakia, (Slovenia), Spain, Switzerland, Turkey.

SYNONYMS: *Elmis coiffaiti* Berthélemy & Clavel, 1961: 243. — Type locality: France. NOTE: Originally, the name *coiffaiti* was blunderingly proposed as a substitute name ("nom. nov.") for "*Helmis fossulata* Sainte-Claire Deville, 1905", however, "*Helmis fossulata* Sainte-Claire Deville, 1905" is not an available name, because Sainte-Claire Deville (1905: 238) did not intend to describe a new taxon since he ascribed the authorship to Kuwert ("*H.* [*Helmis*] *fossulata* Kuw. [Kuwert]"); therefore *Elmis coiffaiti* Berthélemy & Clavel, 1961 has to be regarded as valid description based on "*Helmis fossulata*" (= *Elmis maugetii fossulata* (Kuwert, 1890)) sensu (!) Sainte-Claire Deville (1905); a comprehensive description of *Elmis coiffaiti* was published by Berthélemy (1962: 203); might be a good species.

Elmis minuta Knie, 1975b: 141. — Type locality: Germany.

robusta Jäch, 1984a

Elmis robusta Jäch, 1984a: 136

TYPE LOCALITY: Turkey.

DISTRIBUTION: Iran, Turkey.

syriaca syriaca (Kuwert, 1890)

Lareynia syriaca Kuwert, 1890: 49

TYPE LOCALITY: Lebanon (Berthélemy 1979: 35).

DISTRIBUTION: Israel, Lebanon, Syria.

NOTE: *Elmis syriaca* Kuwert, 1890 was formerly regarded as a junior secondary homonym of *Elmis syriaca* Allard, 1869 (= *Riolus somcheticus* (Kolenati, 1846)) by Zaitzev (1908: 304, 1910: 28), who did not replace this name, which therefore remains valid (ICZN 1999: Art. 59) (see also Berthélemy 1979: 35).

syriaca zoufali (Reitter, 1910)

Helmis zoufali Reitter, 1910: 36

Type Locality: Bosnia and Herzegovina.

Distribution: Albania, Armenia, Bosnia and Herzegovina, Bulgaria, Greece, (Montenegro), Serbia, Turkey.

Elpidelmis Delève, 1964b

Elpidelmis Delève, 1964b: 166. — Type species: *Helmis capensis* Grouvelle, 1890. — Gender feminine.

Note: Incorrect subsequent spelling: "*Epidelmis*" (Bertrand 1972: 484, 504).

capensis (Grouvelle, 1890)

Helmis capensis Grouvelle, 1890: CCXII[212]

Type Locality: South Africa.

Distribution: Angola, South Africa.

fossicollis Delève, 1964b

Elpidelmis fossicollis Delève, 1964b: 169

Type Locality: South Africa.

Distribution: South Africa.

Eonychius Jäch & Boukal, 1996

Eonychius Jäch & Boukal, 1996: 179. — Type species: *Eonychus dudgeoni* Jäch & Boukal, 1995a. — Gender masculine.

Note: Substitute name for *Eonychus* Jäch & Boukal, 1995a: 303.

Homonym: *Eonychus* Jäch & Boukal, 1995a. — Type species: *Eonychus dudgeoni* Jäch & Boukal, 1995a. — Gender masculine. Note: Junior homonym of *Eonychus* Gutierrez, 1969: 44 (Arachnida: Acari).

dudgeoni (Jäch & Boukal, 1995a)

Eonychus dudgeoni Jäch & Boukal, 1995a: 304

Type Locality: China (Hong Kong).

Distribution: China (Fujian, Guangdong, Hong Kong).

Epodelmis Hinton, 1973a

Epodelmis Hinton, 1973a: 5. — Type species: *Epodelmis rosa* Hinton, 1973a. — Gender feminine.

rosa Hinton, 1973a

Epodelmis rosa Hinton, 1973a: 5

TYPE LOCALITY: Bolivia.

DISTRIBUTION: Bolivia.

Esolus Mulsant & Rey, 1872

Esolus Mulsant & Rey, 1872: 36. — Type species: *Limnius parallelepipedus* Müller, 1806a. — Gender masculine.

angustatus (Müller, 1821)

Limnius angustatus Müller, 1821: 187

TYPE LOCALITY: Germany.

DISTRIBUTION: Andorra, Armenia, Austria, Belgium, Bosnia and Herzegovina, Bulgaria, Croatia (Mičetić Stanković et al. 2015: 100), Czech Republic, France, Germany, Greece, Hungary, Italy, Latvia, Luxembourg, Macedonia, (Montenegro), Netherlands, Poland, Portugal, Romania, Russia, Serbia, Slovakia, Slovenia, Spain, Sweden, Switzerland, Turkey, Ukraine.

SYNONYMS: *Esolus carpetanus* Kuwert, 1889: 33. — Type locality: Spain. NOTE: Originally described in a key; authorship ascribed to "Müll. in litt." [Müller in litteris] in original description.

Esolus galloprovincialis Abeille de Perrin, 1900: 137. — Type locality: France.

Esolus solarii Ganglbauer, 1904a: 116. — Type locality: Italy.

berthelemyi Olmi, 1975

Esolus berthelemyi Olmi, 1975: 235

TYPE LOCALITY: Italy.

DISTRIBUTION: France, Italy.

bicuspidatus Alluaud, 1922

Esolus bicuspidatus Alluaud, 1922: 36

TYPE LOCALITY: Morocco.

DISTRIBUTION: Morocco.

brevis Kuwert, 1890

Esolus brevis Kuwert, 1890: 46

TYPE LOCALITY: Italy (Sardinia).

DISTRIBUTION: France (Corsica), Italy (Sardinia).

czwalinae Kuwert, 1889

Esolus czwalinae Kuwert, 1889: 33

TYPE LOCALITY: "Croatia".

DISTRIBUTION: Bosnia and Herzegovina, Croatia, France, Italy, Slovenia.

NOTE: Incorrect subsequent spelling: "*czwalinai*" (Zaitzev 1908: 305, 1910: 29); actual species concept (Berthélemy 1979: 40) based on specimens from Bosnia (Kuwert collection), but not on material from "Croatia".

SYNONYM: *Esolus doderoi* Olmi, 1975: 232. — Type locality: Italy.

filum (Fairmaire, 1871)

Elmis filum Fairmaire, 1871: 373

TYPE LOCALITY: Algeria.

DISTRIBUTION: Algeria, Morocco, Tunisia.

NOTE: *Esolus filum* was erroneously listed as a synonym of *Esolus brevis* Kuwert, 1890 by Mascagni & Calamandrei (1992: 131).

nepalensis Jäch, 1982b

Esolus nepalensis Jäch, 1982b: 91

TYPE LOCALITY: Nepal.

DISTRIBUTION: Nepal.

parallelepipedus (Müller, 1806a)

Limnius parallelepipedus Müller, 1806a: 200

TYPE LOCALITY: Germany.

DISTRIBUTION: Armenia, Austria, Belgium, Bosnia and Herzegovina, Bulgaria, Croatia, Czech Republic, France, Germany, Greece, Hungary, Ireland, Italy, Luxembourg, Macedonia, (Montenegro), Morocco, Netherlands, Poland, Portugal, Romania, (Russia), Serbia, Slovakia, Slovenia, Spain, Switzerland, Turkey, Ukraine, United Kingdom; the record of *E. pygmaeus* from Lithuania by Tamutis et al (2011: 215) and other authors cited therein might well refer to this species.

NOTE: Incorrect subsequent spellings: "*parallelipipedus*" (Sturm 1826: 138, Stephens 1828: 108), "*paralellopipedus*" (Bertolini 1874: 102), "*parallelopipedus*" (Sturm 1857: 17, Kuwert 1889: 33, 1890: 27).

SYNONYMS: *Esolus rugosus* Babington, 1832: 329. — Type locality: United Kingdom (England). NOTE: Incorrect original spelling: "rugo`sus".

Esolus dossowi Kuwert, 1889: 33. — Type locality: France. NOTE: Originally described in a key; formal description published by Kuwert (1890: 47); incorrect subsequent spelling: "*donowi*" (Grouvelle 1896f: 76).

Esolus kuenowi Kuwert, 1889: 33. — Type locality: "Ungarn" (in the late 19th century the territory of "Ungarn" (Hungary) included Croatia (without Dalmatia), Slovakia, large parts of Romania, parts of Serbia and smaller parts of Poland and the Ukraine). NOTE: Incorrect original spelling: "*künowi*"; named after Gotthold Künow (1840–1909), preparator at the

Zoological Museum Königsberg; originally described in a key; formal description published by Kuwert (1890: 47).

Esolus politus Kuwert, 1889: 33. — Type locality: Spain. NOTE: Originally described in a key; authorship ascribed to "Müll. in litt." [Müller in litteris] in original description.

Esolus nicariae Kuwert, 1890: 46. — Type locality: Greece (Ikaria).

Esolus perparvulus Kuwert, 1890: 48. — Type locality: Spain.

pygmaeus (Müller, 1806a)

Limnius pygmaeus Müller, 1806a: 201

TYPE LOCALITY: Germany.

DISTRIBUTION: Algeria, Austria, Belgium, (Bosnia and Herzegovina), Bulgaria, Croatia (Mičetić Stanković et al. 2015: 101), Czech Republic, France, Germany, Greece, Israel, Lebanon, Luxembourg, Morocco, Netherlands, Poland? (Przewoźny et al. 2011), Portugal, Romania, Slovakia, Spain, Switzerland, (Tunisia, record by Boumaiza (1994: 212) unconfirmed), Turkey; the record from Lithuania by Tamutis et al (2011: 215) and other authors cited therein is rather doubtful.

SYNONYMS: *Esolus subparallelus* Kuwert, 1889: 33. — Type locality: France. NOTE: Originally described in a key; authorship ascribed to "Fairm." [Fairmaire] in original description.

Esolus maroccanus Alluaud, 1922: 35. — Type locality: Morocco.

taurus Jäch, 1982b

Esolus taurus Jäch, 1982b: 90

TYPE LOCALITY: Turkey.

DISTRIBUTION: Turkey.

theryi Alluaud, 1922

Esolus theryi Alluaud, 1922: 36

TYPE LOCALITY: Morocco.

DISTRIBUTION: Morocco.

NOTE: Incorrect subsequent spelling: "*théryi*" (Winkler 1926: column 671).

Eumicrodinodes Delève, 1965d

Eumicrodinodes Delève, 1965d: 53. — Type species: *Eumicrodinodes bipustulatus* Delève, 1965d. — Gender masculine.

bipustulatus Delève, 1965d

Eumicrodinodes bipustulatus Delève, 1965d: 54

TYPE LOCALITY: Congo (DR).

DISTRIBUTION: Congo (DR), Congo (R).

concolor Delève, 1967a

Eumicrodinodes concolor Delève, 1967a: 330

TYPE LOCALITY: Congo (R).

DISTRIBUTION: Congo (R).

quadrimaculatus Delève, 1972

Eumicrodinodes quadrimaculatus Delève, 1972: 920

TYPE LOCALITY: Gabon.

DISTRIBUTION: Gabon.

Exolimnius Delève, 1954

Exolimnius Delève, 1954: 27. — Type species: *Exolimnius ungulatus* Delève, 1954. — Gender masculine.

lateritius (Fairmaire, 1902)

Elmidolia lateritia Fairmaire, 1902: 344 ("*Elmidola*")

TYPE LOCALITY: Madagascar.

DISTRIBUTION: Madagascar.

NOTE: Incorrect subsequent spelling: "*Exolimnius lateritia*" (Delève 1963d: 8), the epithet name lateritius (= brick-red) is an adjective.

ungulatus Delève, 1954

Exolimnius ungulatus Delève, 1954: 28

TYPE LOCALITY: Madagascar.

DISTRIBUTION: Madagascar.

Gonielmis Sanderson, 1953a

Gonielmis Sanderson, 1953a: 155. — Type species: *Elmis dietrichi* Musgrave, 1933. — Gender feminine.

NOTE: Originally described in a key; formal description published by Sanderson (1954: 5); incorrect subsequent spelling: "*Gonioelmis*" (Bertrand 1972: 501).

dietrichi (Musgrave, 1933)

Helmis dietrichi Musgrave, 1933: 54

TYPE LOCALITY: USA (Mississippi).

DISTRIBUTION: USA (Alabama, Florida, Georgia, Louisiana, Mississippi, South Carolina, Tennessee).

NOTE: Incorrect subsequent spelling: "*dietrechi*" (Bertrand 1972: 501).

Graphelmis Delève, 1968c

Graphelmis Delève, 1968c: 169. — Type species: *Graphelmis scapularis* Delève, 1968c. — Gender feminine.

NOTE: Incorrect subsequent spelling: "*Grphelmis*" (Čiampor 2001: 20).

ambigua Delève, 1970d

Graphelmis ambigua Delève, 1970d: 260

TYPE LOCALITY: Vietnam.

DISTRIBUTION: Thailand, Vietnam.

anulata Čiampor, 2006

Graphelmis anulata Čiampor, 2006: 13

TYPE LOCALITY: Malaysia (Pahang).

DISTRIBUTION: Malaysia (Pahang).

balkei Čiampor & Kodada, 2004

balkei Čiampor & Kodada, 2004: 83

TYPE LOCALITY: Indonesia (Papua).

DISTRIBUTION: Indonesia (Papua), Papua New Guinea.

bandukanensis Čiampor, 2002a

Graphelmis bandukanensis Čiampor, 2002a: 149

TYPE LOCALITY: Malaysia (Sabah).

DISTRIBUTION: Malaysia (Sabah, Sarawak).

basalis Čiampor & Kodada, 2004

Graphelmis basalis Čiampor & Kodada, 2004: 68

TYPE LOCALITY: Papua New Guinea.

DISTRIBUTION: Papua New Guinea.

berbulu Čiampor, 2002a

Graphelmis berbulu Čiampor, 2002a: 155

TYPE LOCALITY: Malaysia (Sarawak).

DISTRIBUTION: Malaysia (Sarawak).

binervosa (Reitter, 1887)

Stenelmis binervosa Reitter, 1887: 259

TYPE LOCALITY: Papua New Guinea.

DISTRIBUTION: Papua New Guinea.

NOTE: Incorrect original spelling: "*Stenelmis binervosus*".

biroi (Bollow, 1942)

Stenelmis biroi Bollow, 1942: 197

TYPE LOCALITY: Papua New Guinea.

DISTRIBUTION: Papua New Guinea.

NOTE: Incorrect original spelling: "*biróï*".

bouchardi (Grouvelle, 1896d)

Cylloepus bouchardi Grouvelle, 1896d: 44

TYPE LOCALITY: Indonesia (Sumatra).

DISTRIBUTION: Indonesia (Sumatra).

boukali Čiampor, 2004b

Graphelmis boukali Čiampor, 2004b: 8

TYPE LOCALITY: Thailand.

DISTRIBUTION: Laos, Malaysia (Kelantan, Pahang), Thailand.

brezanskae Čiampor, 2003

Graphelmis brezanskae Čiampor, 2003: 39

TYPE LOCALITY: Malaysia (Sarawak).

DISTRIBUTION: Malaysia (Sarawak).

bruneica Čiampor & Kodada, 2004

Graphelmis bruneica Čiampor & Kodada, 2004: 62

TYPE LOCALITY: Brunei.

DISTRIBUTION: Brunei.

ceylonica (Motschulsky, 1860)

Stenelmis ceylonica Motschulsky, 1860: 49

TYPE LOCALITY: Sri Lanka.

DISTRIBUTION: Sri Lanka.

NOTE: Incorrect original spelling: "*Stenelmis ceylonicus*".

clava Čiampor & Kodada, 2004

Graphelmis clava Čiampor & Kodada, 2004: 85

TYPE LOCALITY: Indonesia (Papua).

DISTRIBUTION: Indonesia (Papua).

clermonti (Pic, 1923)

Stenelmis clermonti Pic, 1923: 4

TYPE LOCALITY: Vietnam.

DISTRIBUTION: China (Yunnan), Laos, Thailand, Vietnam.

consobrina Delève, 1968c

Graphelmis consobrina Delève, 1968c: 173

TYPE LOCALITY: Vietnam.

DISTRIBUTION: Vietnam.

convexa Čiampor & Kodada, 2004

Graphelmis convexa Čiampor & Kodada, 2004: 72

TYPE LOCALITY: Indonesia (Papua).

DISTRIBUTION: Indonesia (Papua).

cornuta Čiampor, 2004b

Graphelmis cornuta Čiampor, 2004b: 14

TYPE LOCALITY: Vietnam.

DISTRIBUTION: Vietnam.

darwini Čiampor & Kodada, 2004

Graphelmis darwini Čiampor & Kodada, 2004: 67

TYPE LOCALITY: Indonesia (Borneo).

DISTRIBUTION: Indonesia (Borneo), Malaysia (Sarawak).

delevei Čiampor, 2001

Graphelmis delevei Čiampor, 2001: 27

TYPE LOCALITY: China (Sichuan).

DISTRIBUTION: China (Sichuan).

dembickyi Čiampor, 2002a

Graphelmis dembickyi Čiampor, 2002a: 157

TYPE LOCALITY: Philippines (Luzon).

DISTRIBUTION: Philippines (Luzon, Mindoro).

dentipes Čiampor, 2005a

Graphelmis dentipes Čiampor, 2005a: 16

TYPE LOCALITY: Malaysia (Sabah).

DISTRIBUTION: Malaysia (Sabah).

diaphoroderes Jäch, 1985

Graphelmis diaphoroderes Jäch, 1985: 237

TYPE LOCALITY: Papua New Guinea.

DISTRIBUTION: Papua New Guinea.

elegans Čiampor, 2005a

Graphelmis elegans Čiampor, 2005a: 30

TYPE LOCALITY: Laos.

DISTRIBUTION: Laos.

elisabethjaechae Čiampor, 2005a

Graphelmis elisabethjaechae Čiampor, 2005a: 33

TYPE LOCALITY: Philippines (Luzon).

DISTRIBUTION: Philippines (Luzon).

fulvolineata Čiampor, 2004b

Graphelmis fulvolineata Čiampor, 2004b: 10

TYPE LOCALITY: Laos.

DISTRIBUTION: Laos, Vietnam.

fuscicornis Čiampor, 2005a

Graphelmis fuscicornis Čiampor, 2005a: 18

TYPE LOCALITY: Vietnam.

DISTRIBUTION: Vietnam.

gemuk Čiampor, 2005b

Graphelmis gemuk Čiampor, 2005b: 120

TYPE LOCALITY: Malaysia (Sarawak).

DISTRIBUTION: Malaysia (Sarawak).

gibberosa Čiampor & Kodada, 2004

Graphelmis gibberosa Čiampor & Kodada, 2004: 63

TYPE LOCALITY: Indonesia (Borneo).

DISTRIBUTION: Indonesia (Borneo).

grouvellei Delève, 1970d

Graphelmis grouvellei Delève, 1970d: 251

TYPE LOCALITY: Indonesia (Sumatra).

DISTRIBUTION: Indonesia (Borneo, Sumatra), Malaysia (Pahang).

hintoni Čiampor, 2006

Graphelmis hintoni Čiampor, 2006: 16

Type Locality: Malaysia (Sabah).

Distribution: Malaysia (Sabah); this species does not occur in Pahang (Malaysia), the label data of the type specimens published in the original description were incorrectly cited.

hlavaci Čiampor & Kodada, 2004

Graphelmis hlavaci Čiampor & Kodada, 2004: 71

Type Locality: Papua New Guinea.

Distribution: Papua New Guinea.

insolita Čiampor & Kodada, 2004

Graphelmis insolita Čiampor & Kodada, 2004: 95

Type Locality: Indonesia (Papua).

Distribution: Indonesia (Papua).

jaechi Čiampor, 2001

Graphelmis jaechi Čiampor, 2001: 20 ("*Grphelmis*")

Type Locality: China (Hong Kong).

Distribution: China (Anhui, Fujian, Guangdong, Guangxi, Hong Kong, Hunan, Jiangxi, Yunnan), Laos, Thailand, Vietnam.

jendeki Čiampor & Kodada, 2004

Graphelmis jendeki Čiampor & Kodada, 2004: 63

Type Locality: Thailand.

Distribution: Laos, Thailand.

jesusi Čiampor & Kodada, 2004

Graphelmis jesusi Čiampor & Kodada, 2004: 100

Type Locality: Indonesia (Papua).

Distribution: Indonesia (Papua).

kodadai Čiampor, 2003

Graphelmis kodadai Čiampor, 2003: 40

Type Locality: Malaysia (Sabah).

Distribution: Malaysia (Sabah).

kuamutensis Čiampor, 2003

Graphelmis kuamutensis Čiampor, 2003: 35

Type Locality: Malaysia (Sabah).

Distribution: Malaysia (Sabah).

kubani Čiampor & Kodada, 2004

Graphelmis kubani Čiampor & Kodada, 2004: 59

Type Locality: Vietnam.

Distribution: Vietnam.

labralis Čiampor, 2003

Graphelmis labralis Čiampor, 2003: 31

Type Locality: Malaysia (Sabah).

Distribution: Malaysia (Sabah).

loebli Čiampor & Kodada, 2004

Graphelmis loebli Čiampor & Kodada, 2004: 79

Type Locality: Indonesia (Papua).

Distribution: Indonesia (Papua).

lomata Jäch, 1985

Graphelmis lomata Jäch, 1985: 236

Type Locality: Papua New Guinea.

Distribution: Papua New Guinea.

malickyi Čiampor, 2005a

Graphelmis malickyi Čiampor, 2005a: 23

Type Locality: Indonesia (Sumatra).

Distribution: Indonesia (Sumatra).

marshalli (Hinton, 1936e)

Cylloepus marshalli Hinton, 1936e: 220

Type Locality: Malaysia (Sarawak).

Distribution: Malaysia (Sabah, Sarawak).

merkli Čiampor, 2006

Graphelmis merkli Čiampor, 2006: 16

Type Locality: Indonesia (Borneo).

Distribution: Indonesia (Borneo).

minuta Čiampor, 2002a

Graphelmis minuta Čiampor, 2002a: 154

Type Locality: Malaysia (Sarawak).

Distribution: Malaysia (Sarawak).

monticola (Grouvelle, 1896d)

Helmis monticola Grouvelle, 1896d: 45

Type Locality: Indonesia (Sumatra).

Distribution: Indonesia (Sumatra), Malaysia (Kedah, Kelantan, Pahang, Penang, Perak, Selangor).

mumini Čiampor, 2001

Graphelmis mumini Čiampor, 2001: 31

Type Locality: Malaysia (Sarawak).

Distribution: Malaysia (Sarawak).

nigromaculata (Chûjô & Satô, 1964) **comb.n.**

Stenelmis nigromaculata Chûjô & Satô, 1964: 195

Type Locality: Malaysia (Sarawak).

Distribution: Malaysia (Sarawak).

Note: This species is transferred herewith from *Stenelmis* Dufour, 1835; based on the original description (Chûjô & Satô 1964: Fig. 8) it strongly resembles *Graphelmis marshalli* (Hinton, 1936e); holotype not retrieved, not in Ehime University, Matsuyama, and not in Kyushu University Museum, Fukuoka, it might be deposited in Kyoto University Museum.

nitida Čiampor, 2005a

Graphelmis nitida Čiampor, 2005a: 26

Type Locality: Malaysia (Sarawak).

Distribution: Malaysia (Sarawak).

obesa Čiampor, 2005b

Graphelmis obesa Čiampor, 2005b: 117

Type Locality: Malaysia (Sabah).

Distribution: Malaysia (Sabah).

oxytela Jäch, 1985

Graphelmis oxytela Jäch, 1985: 233

Type Locality: Papua New Guinea.

Distribution: Papua New Guinea.

palawanensis Čiampor & Kodada, 2004

Graphelmis palawanensis Čiampor & Kodada, 2004: 60

Type Locality: Philippines (Palawan).

Distribution: Philippines (Palawan).

pallidipes (Carter, 1926b)

Helmis pallidipes Carter, 1926b: 63

Type Locality: Australia (New South Wales).

Distribution: Australia (New South Wales, Queensland).

philemoni Čiampor & Kodada, 2004

Graphelmis philemoni Čiampor & Kodada, 2004: 65

Type Locality: Malaysia (Sarawak).

Distribution: Indonesia (Borneo), Malaysia (Sabah, Sarawak).

picea Čiampor & Kodada, 2004

Graphelmis picea Čiampor & Kodada, 2004: 87

Type Locality: Indonesia (Papua).

Distribution: Indonesia (Papua).

picta (Reitter, 1886)

Stenelmis picta Reitter, 1886: 213

Type Locality: Indonesia (Sumatra).

Distribution: Indonesia (Siberut, Sumatra), Thailand?

Note: Incorrect original spelling: "*Stenelmis pictus*".

posoica Čiampor & Kodada, 2004

Graphelmis posoica Čiampor & Kodada, 2004: 92

Type Locality: Indonesia (Sulawesi).

Distribution: Indonesia (Sulawesi, Togian Islands).

prisca Čiampor, 2000

Graphelmis prisca Čiampor, 2000: 183

Type Locality: India (Kerala).

Distribution: India (Kerala).

punggulensis Čiampor, 2002a

Graphelmis punggulensis Čiampor, 2002a: 152

Type Locality: Malaysia (Sabah).

Distribution: Brunei, Malaysia (Sabah, Sarawak).

quadrimaculata Čiampor, 2005a

Graphelmis quadrimaculata Čiampor, 2005a: 15

Type Locality: Malaysia (Sarawak).

Distribution: Malaysia (Pahang, Sarawak).

reitteri Čiampor & Kodada, 2004

Graphelmis reitteri Čiampor & Kodada, 2004: 97

TYPE LOCALITY: Indonesia (Papua).

DISTRIBUTION: Indonesia (Papua).

riberai Čiampor & Kodada, 2004

Graphelmis riberai Čiampor & Kodada, 2004: 86

TYPE LOCALITY: Indonesia (Papua).

DISTRIBUTION: Indonesia (Papua).

robusta Čiampor, 2005a

Graphelmis robusta Čiampor, 2005a: 21

TYPE LOCALITY: Indonesia (Java).

DISTRIBUTION: Indonesia (Java).

sausai Čiampor, 2004b

Graphelmis sausai Čiampor, 2004b: 17

TYPE LOCALITY: Thailand.

DISTRIBUTION: Laos, Thailand.

scapularis Delève, 1968c

Graphelmis scapularis Delève, 1968c: 171

TYPE LOCALITY: Vietnam.

DISTRIBUTION: Vietnam.

schillhammeri Čiampor, 2004b

Graphelmis schillhammeri Čiampor, 2004b: 12

TYPE LOCALITY: Laos.

DISTRIBUTION: China (Yunnan), Laos.

schneideri Čiampor & Kodada, 2004

Graphelmis schneideri Čiampor & Kodada, 2004: 76

TYPE LOCALITY: Indonesia (Borneo).

DISTRIBUTION: Indonesia (Borneo).

schoedli Čiampor, 2005a

Graphelmis schoedli Čiampor, 2005a: 33

TYPE LOCALITY: Philippines (Mindoro).

DISTRIBUTION: Philippines (Mindoro).

securipes Čiampor & Kodada, 2004

Graphelmis securipes Čiampor & Kodada, 2004: 74

Type Locality: Indonesia (Papua).

Distribution: Indonesia (Papua).

shirahatai (Nomura, 1958a)

Stenelmis shirahatai Nomura, 1958a: 42

Type Locality: Japan (Honshu).

Distribution: Japan (Honshu).

sniffi Čiampor, 2001

Graphelmis sniffi Čiampor, 2001: 30

Type Locality: Malaysia (Sabah).

Distribution: Malaysia (Sabah).

spiralis Čiampor, 2004b

Graphelmis spiralis Čiampor, 2004b: 14

Type Locality: Thailand.

Distribution: Thailand.

strbaki Čiampor & Kodada, 2004

Graphelmis strbaki Čiampor & Kodada, 2004: 95

Type Locality: Indonesia (Papua).

Distribution: Indonesia (Papua).

sulawesiensis Čiampor, 2002a

Graphelmis sulawesiensis Čiampor, 2002a: 159

Type Locality: Indonesia (Sulawesi).

Distribution: Indonesia (Sulawesi).

tawauica Čiampor, 2003

Graphelmis tawauica Čiampor, 2003: 33

Type Locality: Malaysia (Sabah).

Distribution: Malaysia (Sabah).

temburongensis Čiampor, 2003

Graphelmis temburongensis Čiampor, 2003: 43

Type Locality: Brunei.

Distribution: Brunei.

togianica Čiampor & Kodada, 2004

Graphelmis togianica Čiampor & Kodada, 2004: 98

TYPE LOCALITY: Indonesia (Togian Islands).

DISTRIBUTION: Indonesia (Sulawesi, Togian Islands).

tuberculata Čiampor & Kodada, 2004

Graphelmis tuberculata Čiampor & Kodada, 2004: 72

TYPE LOCALITY: Indonesia (Papua).

DISTRIBUTION: Indonesia (Papua).

valida Delève, 1970d

Graphelmis valida Delève, 1970d: 258

TYPE LOCALITY: Indonesia (Sumatra).

DISTRIBUTION: Indonesia (Sumatra).

verpa Jäch, 1985

Graphelmis verpa Jäch, 1985: 239

TYPE LOCALITY: Papua New Guinea.

DISTRIBUTION: Papua New Guinea.

NOTE: The epithet name is a noun in apposition (penis).

vietnamensis Čiampor, 2005a

Graphelmis vietnamensis Čiampor, 2005a: 25

TYPE LOCALITY: Vietnam.

DISTRIBUTION: Vietnam.

vogleri Čiampor & Kodada, 2004

Graphelmis vogleri Čiampor & Kodada, 2004: 82

TYPE LOCALITY: Indonesia (Papua).

DISTRIBUTION: Indonesia (Papua).

Graphosolus Jäch & Kodada, 1996a

Graphosolus Jäch & Kodada, 1996a: 93. — Type species: *Graphosolus philippinensis* Jäch & Kodada, 1996a. — Gender masculine.

javanicus Jäch & Kodada, 1996a

Graphosolus javanicus Jäch & Kodada, 1996a: 98

TYPE LOCALITY: Indonesia (Java).

DISTRIBUTION: Indonesia (Java).

philippinensis Jäch & Kodada, 1996a

Graphosolus philippinensis Jäch & Kodada, 1996a: 94

Type Locality: Philippines (Luzon).

Distribution: Philippines (Luzon, Mindoro, Negros).

tiomani Čiampor, 2004a

Graphosolus tiomani Čiampor, 2004a: 222

Type Locality: Malaysia (Johor).

Distribution: Malaysia (Johor).

variabilis palawanensis Jäch & Kodada, 1996a

Graphosolus variabilis palawanensis Jäch & Kodada, 1996a: 98

Type Locality: Philippines (Palawan).

Distribution: Philippines (Palawan).

variabilis variabilis Jäch & Kodada, 1996a

Graphosolus variabilis Jäch & Kodada, 1996a: 94

Type Locality: Malaysia (Sarawak).

Distribution: Indonesia (Sumatra), Malaysia (Sarawak).

Grouvellinus Champion, 1923

Grouvellinus Champion, 1923: 168. — Type species: *Macronychus caucasicus* Victor, 1839. — Gender masculine.

Note: Substitute name for *Microdes* Motschulsky, 1860; incorrect subsequent spelling: "*Grouvelinus*" (Hua 2002: 98).

Homonyms: *Microdes* Motschulsky, 1860: 48. — Type species: *Macronychus caucasicus* Victor, 1839. — Gender masculine (ICZN 1999: Art. 30.1.4.4). Note: Junior homonym of *Microdes* Guenée, 1857: 296 (Lepidoptera).

Grouvelleus Zaitzev, 1908: 311. — Type species: *Macronychus caucasicus* Victor, 1839. — Gender masculine. Note: Substitute name for *Microdes* Motschulsky, 1860; junior homonym of *Grouvelleus* Guillebeau, 1892: CXXXIV[134] (Phalacridae).

aeneus (Grouvelle, 1896d)

Microdes aeneus Grouvelle, 1896d: 47

Type Locality: Indonesia (Sumatra).

Distribution: Indonesia (Bali?, Sumatra).

amabilis Delève, 1970d

Grouvellinus amabilis Delève, 1970d: 263

Type Locality: Vietnam.

Distribution: Vietnam.

babai babai Nomura, 1963

Grouvellinus babai Nomura, 1963: 54

Type Locality: Taiwan.

Distribution: Taiwan.

babai satoi Jeng & Yang, 1998

Grouvellinus babai satoi Jeng & Yang, 1998: 538

Type Locality: Japan (Ryukyu Islands).

Distribution: Japan (Ryukyu Islands).

bishopi Jäch, 1984c

Grouvellinus bishopi Jäch, 1984c: 123

Type Locality: Malaysia (Selangor).

Distribution: Malaysia (Selangor).

brevior Jäch, 1984c

Grouvellinus brevior Jäch, 1984c: 110

Type Locality: Nepal.

Distribution: Nepal.

carinatus Jäch, 1984c

Grouvellinus carinatus Jäch, 1984c: 111

Type Locality: Nepal.

Distribution: Nepal.

carus Hinton, 1941b

Grouvellinus carus Hinton, 1941b: 69

Type Locality: China (Fujian).

Distribution: China (Fujian).

caucasicus (Victor, 1839)

Macronychus caucasicus Victor, 1839: 70

Type Locality: Georgia.

Distribution: Armenia, Georgia, Greece (Samos), Iran, Iraq, Israel, Lebanon, Russia, Syria, Turkey.

Note: This species was published by V.I. Motschulsky under the pseudonym T. Victor; authorship incorrectly ascribed to Motschulsky by all subsequent authors.

Synonym: *Elmis coyei* Allard, 1869: 466. — Type locality: Lebanon.

chinensis Mařan, 1939

Grouvellinus chinensis Mařan, 1939: 42

Type Locality: China (Sichuan).

Distribution: China (Sichuan).

duplaris Champion, 1923

Grouvellinus duplaris Champion, 1923: 168

Type Locality: India (Uttarakhand).

Distribution: India (Himachal Pradesh, Uttarakhand).

frater (Grouvelle, 1896d)

Microdes frater Grouvelle, 1896d: 46

Type Locality: Indonesia (Sumatra).

Distribution: Indonesia (Sumatra).

hadroscelis Jäch, 1984c

Grouvellinus hadroscelis Jäch, 1984c: 115

Type Locality: Nepal.

Distribution: Nepal.

hercules Jäch, 1984c

Grouvellinus hercules Jäch, 1984c: 113

Type Locality: Nepal.

Distribution: China (Tibet) Nepal.

hygropetricus Jeng & Yang, 1998

Grouvellinus hygropetricus Jeng & Yang, 1998: 533

Type Locality: Taiwan.

Distribution: Taiwan.

impressus Jäch, 1984c

Grouvellinus impressus Jäch, 1984c: 122

Type Locality: Indonesia (Sumatra).

Distribution: Indonesia (Sumatra).

marginatus (Kôno, 1934)

Grouvelleus marginatus Kôno, 1934: 127

Type Locality: Japan (Shikoku).

Distribution: Japan (Honshu, Kyushu, Shikoku, Yakushima).

modiglianii (Grouvelle, 1896d)

Microdes modiglianii Grouvelle, 1896d: 46

Type Locality: Indonesia (Sumatra).

Distribution: Indonesia (Sumatra).

montanus Jeng & Yang, 1998

Grouvellinus montanus Jeng & Yang, 1998: 530

Type Locality: Taiwan.

Distribution: Taiwan.

nepalensis Delève, 1970f

Grouvellinus nepalensis Delève, 1970f: 319

Type Locality: Nepal.

Distribution: China (Tibet), Nepal.

nitidus Nomura, 1963

Grouvellinus nitidus Nomura, 1963: 52

Type Locality: Japan (Honshu).

Distribution: Japan (Honshu).

pelacoti Delève, 1970d

Grouvellinus pelacoti Delève, 1970d: 268

Type Locality: Vietnam.

Distribution: Vietnam.

pilosus Jeng & Yang, 1998

Grouvellinus pilosus Jeng & Yang, 1998: 527

Type Locality: Taiwan.

Distribution: Taiwan.

punctatostriatus Bollow, 1940b

Grouvellinus punctatostriatus Bollow, 1940b: 32

Type Locality: Myanmar.

Distribution: Myanmar.

rioloides (Reitter, 1887)

Macronychus rioloides Reitter, 1887: 259

Type Locality: Kazakhstan; there has been some confusion about the Type locality: Reitter (1887: 260) wrote: "Turkestan (Fluß Tamga). Von König und Faust eingesendet"; Tamga River is in Kyrgystan; however the label of the lectotype, deposited in the Hungarian

Natural History Museum, says: "Turkestan, Aulie Fluß [= river] Tanga [= Tamga] leg. König", which is a mixture of two different locations: "Aulie" [= Taraz] is a town in Kazakhstan and Tamga River refers to Kyrgystan; in any case, this is not an original label, it was added later by one of the museum curators in the Hungarian Natural History Museum; fortunately, underneath of this label there is a small grey original label saying: "Aulie", which provides clear evidence of the true type locality.

DISTRIBUTION: Afghanistan, China (Xinjiang), Kazakhstan, Kyrgystan, Tajikistan, Turkmenistan, Uzbekistan.

SYNONYMS: *Macronychus arius* Janssens, 1959a: 3. — Type locality: Afghanistan.

Macronychus rioloides flavibasis Reitter, 1887: 260. — Type locality: Kazakhstan. NOTE: Originally described as a variety of *Macronychus rioloides*.

sculptus Bollow, 1940b

Grouvellinus sculptus Bollow, 1940b: 35

TYPE LOCALITY: Myanmar.

DISTRIBUTION: Myanmar.

setosus Delève, 1970d

Grouvellinus setosus Delève, 1970d: 266

TYPE LOCALITY: Vietnam.

DISTRIBUTION: Vietnam.

silius Hinton, 1941b

Grouvellinus silius Hinton, 1941b: 71

TYPE LOCALITY: Indonesia (Java).

DISTRIBUTION: Indonesia (Java).

NOTE: This species could be a junior synonym of *Grouvellinus aeneus* (Grouvelle, 1896d)

sinensis (Grouvelle, 1906b)

Microdes sinensis Grouvelle, 1906b: 125

TYPE LOCALITY: China (Yunnan).

DISTRIBUTION: China (Yunnan).

subopacus Nomura, 1962

Grouvellinus subopacus Nomura, 1962: 48

TYPE LOCALITY: Japan (Ryukyu Islands).

DISTRIBUTION: Japan (Ryukyu Islands).

sumatrensis Jäch, 1984c

Grouvellinus sumatrensis Jäch, 1984c: 120

Type Locality: Indonesia (Sumatra).

Distribution: Indonesia (Sumatra).

thienemanni Jäch, 1984c

Grouvellinus thienemanni Jäch, 1984c: 119

Type Locality: Indonesia (Java).

Distribution: Indonesia (Java).

tibetanus Jäch, 1984c

Grouvellinus tibetanus Jäch, 1984c: 114

Type Locality: Nepal.

Distribution: China (Tibet), Nepal.

tonkinus (Grouvelle, 1889b)

Microdes tonkinus Grouvelle, 1889b: 409

Type Locality: Vietnam.

Distribution: Vietnam.

unicostatus Champion, 1923

Grouvellinus unicostatus Champion, 1923: 169

Type Locality: India (Uttarakhand).

Distribution: India (Himachal Pradesh, Uttarakhand).

Gyrelmis Hinton, 1940b

Gyrelmis Hinton, 1940b: 381. — Type species: *Gyrelmis brunnea* Hinton, 1940b. — Gender feminine.

brunnea Hinton, 1940b

Gyrelmis brunnea Hinton, 1940b: 405

Type Locality: Brazil (Pará).

Distribution: Brazil (Goiás, Pará), French Guiana.

glabra Hinton, 1940b

Gyrelmis glabra Hinton, 1940b: 388

Type Locality: French Guiana.

Distribution: Brazil (Amazonas), French Guiana.

longipes Hinton, 1940b

Gyrelmis longipes Hinton, 1940b: 400

Type Locality: Brazil (Pará).

Distribution: Brazil (Pará, Rondônia), French Guiana; erroneously recorded from the Brazilian State of Mato Grosso by Hinton (1940b: 402) and Segura et al. (2013: 17).

maculata Hinton, 1940b

Gyrelmis maculata Hinton, 1940b: 402

Type Locality: Brazil (Pará).

Distribution: Brazil (Amazonas, Pará).

nubila Hinton, 1940b

Gyrelmis nubila Hinton, 1940b: 407

type Locality: Brazil (Rondônia); erroneously, Hinton (1940b: 409) attributed the type locality to the State of Mato Grosso.

Distribution: Brazil (Rondônia); erroneously recorded from Mato Grosso by Segura et al. (2013: 17).

obesa Hinton, 1940b

Gyrelmis obesa Hinton, 1940b: 393

Type Locality: French Guiana.

Distribution: Brazil (Amazonas), French Guiana.

pulchella Hinton, 1940b

Gyrelmis pulchella Hinton, 1940b: 406

Type Locality: French Guiana.

Distribution: French Guiana.

pusio Hinton, 1940b

Gyrelmis pusio Hinton, 1940b: 398

Type Locality: Brazil (Rondônia); erroneously, Hinton (1940b: 400) attributed the type locality to the State of Mato Grosso.

Distribution: Brazil (Amazonas, Rondônia); erroneously recorded from Mato Grosso by Passos et al. (2010a: 540) and Segura et al. (2013: 18).

rufomarginata (Grouvelle, 1889b)

Helmis rufomarginata Grouvelle, 1889b: 405

Type Locality: Brazil (Santa Catarina).

Distribution: Brazil (Goiás, Santa Catarina).

Note: Incorrect original spelling: "*Helmis rufo-marginata*".

simplex Hinton, 1940b

Gyrelmis simplex Hinton, 1940b: 392

TYPE LOCALITY: Brazil (Amazonas).

DISTRIBUTION: Brazil (Amazonas, Pará, Rondônia), French Guiana; erroneously recorded from the Brazilian State of Mato Grosso by Hinton (1940b: 392) and Segura et al. (2013: 18).

spinata Hinton, 1940b

Gyrelmis spinata Hinton, 1940b: 390

TYPE LOCALITY: French Guiana.

DISTRIBUTION: Brazil (Amazonas), French Guiana.

thoracica basalis Hinton, 1940b

Gyrelmis thoracica basalis Hinton, 1940b: 398

TYPE LOCALITY: Brazil (Amazonas).

DISTRIBUTION: Brazil (Amazonas, Pará, Rondônia); erroneously recorded from French Guiana (Passos et al. 2010a: 540, Segura et al. 2013: 18) and the Brazilian State of Mato Grosso (Hinton 1940b: 398, Segura et al. 2013: 18).

thoracica thoracica Hinton, 1940b

Gyrelmis thoracica Hinton, 1940b: 397

TYPE LOCALITY: French Guiana.

DISTRIBUTION: French Guiana.

Haraldaria Jäch & Boukal, 1996

Haraldaria Jäch & Boukal, 1996: 183. — Type species: *Haraldaria schillhammeri* Jäch & Boukal, 1996. — Gender feminine.

schillhammeri Jäch & Boukal, 1996

Haraldaria schillhammeri Jäch & Boukal, 1996: 189

TYPE LOCALITY: Malaysia (Pahang).

DISTRIBUTION: Malaysia (Pahang).

Hedyselmis Hinton, 1976

Hedyselmis Hinton, 1976: 259. — Type species: *Hedyselmis opis* Hinton, 1976. — Gender feminine.

belatani Čiampor & Čiamporová-Zaťovičová, 2008

Hedyselmis belatani Čiampor & Čiamporová-Zaťovičová, 2008: 56

TYPE LOCALITY: Malaysia (Terengganu).

DISTRIBUTION: Malaysia (Terengganu).

gibbosa Jäch & Boukal, 1997b

Hedyselmis gibbosa Jäch & Boukal, 1997b: 116

Type Locality: Malaysia (Perak).

Distribution: Malaysia (Perak, Kelantan).

Note: Incorrect original spelling: "*gibbosus*", herewith changed mandatorily (ICZN 1999: Art. 34.2).

opis Hinton, 1976

Hedyselmis opis Hinton, 1976: 259

Type Locality: Malaysia (Pahang).

Distribution: Malaysia (Pahang).

Helminthocharis Grouvelle, 1906c

Helminthocharis Grouvelle, 1906c: 321. — Type species: *Helminthocharis picea* Grouvelle, 1906c. — Gender feminine.

Note: The type species is herewith designated; according to Delève (1964a: 45) the genus *Helminthocharis* was "établi sur une espèce (*H. picea* Grouvelle)", but in fact it was established on two species, *Helminthocharis picea* Grouvelle, 1906c and *Elmis nitidula* Fairmaire, 1897b (= *Helminthocharis filicornis* Jäch & Kodada nom.n.), and therefore this type designation is necessary (ICZN 1999: Art. 67); incorrect subsequent spelling: "*Helminthocaris*" (Delève 1938: 372).

abdominalis abdominalis Delève, 1956

Helminthocharis abdominalis Delève, 1956: 378

Type Locality: Rwanda.

Distribution: Angola, Central African Republic, Ghana, Rwanda.

abdominalis nigra Delève, 1967c

Helminthocharis abdominalis nigra Delève, 1967c: 78

Type Locality: Congo (R).

Distribution: Congo (R), Ghana.

congoensis Delève, 1967c

Helminthocharis congoensis Delève, 1967c: 79

Type Locality: Congo (R).

Distribution: Congo (R).

cristula Delève, 1967b

Helminthocharis cristula Delève, 1967b: 440

Type Locality: South Africa.

Distribution: South Africa.

diasticta Alluaud, 1933

Helminthocharis diasticta Alluaud, 1933: 157

Type Locality: Ivory Coast.

Distribution: Angola, Burkina Faso (Delève 1966d: 59), Central African Republic, Congo (DR) (Delève 1955: 20, but not included in Delève 1966d: 59), Ivory Coast.

filicornis Jäch & Kodada, 2016 **nom.n.**

Helminthocharis filicornis Jäch & Kodada, 2016: XVIII

Type Locality: Madagascar.

Distribution: Madagascar.

Note: First published as a nomen nudum by Delève (1964a: 47); substitute name for *Elmis nitidula* Fairmaire, 1897b.

Homonym: *Elmis nitidula* Fairmaire, 1897b: 369. Note: Incorrect original spelling: "*nitidulus*"; junior primary homonym of *Elmis nitidula* Leconte, 1866 (*Oulimnius*); transferred to *Helmis* by Grouvelle (1906c: 321).

picea Grouvelle, 1906c

Helminthocharis picea Grouvelle, 1906c: 321

Type Locality: Tanzania.

Distribution: Congo (DR) (Delève 1938: 372), Tanzania.

polita Delève, 1964a

Helminthocharis polita Delève, 1964a: 47

Type Locality: Madagascar.

Distribution: Madagascar.

schoutedeni Delève, 1938

Helminthocharis schoutedeni Delève, 1938: 372

Type Locality: Congo (DR).

Distribution: Angola, Central African Republic, Congo (DR), Ghana, Liberia.

Helminthopsis Grouvelle, 1906c

Helminthopsis Grouvelle, 1906c: 319. — Type species: *Helminthopsis lucida* Grouvelle, 1906c. — Gender feminine.

Note: The type species is herewith designated; this genus might be a junior synonym of *Ludyella* Reitter, 1899.

SUBGENUS: ***Elmidoliana*** Delève, 1965b: 22. — Type species: *Helminthopsis luteopicta* Delève, 1938. — Gender feminine.

allansoni Delève, 1967b

Helminthopsis allansoni Delève, 1967b: 438

TYPE LOCALITY: South Africa.

DISTRIBUTION: South Africa.

ambigua Delève, 1974

Helminthopsis ambigua Delève, 1974: 277 (subgenus *Elmidoliana*)

TYPE LOCALITY: Ghana.

DISTRIBUTION: Ghana.

assimilis Delève, 1967a

Helminthopsis assimilis Delève, 1967a: 333

TYPE LOCALITY: Congo (R).

DISTRIBUTION: Congo (R).

bifida Delève, 1965b

Helminthopsis bifida Delève, 1965b: 16

TYPE LOCALITY: South Africa.

DISTRIBUTION: South Africa, Zimbabwe.

castanea Delève, 1965b

Helminthopsis castanea Delève, 1965b: 10

TYPE LOCALITY: Congo (DR).

DISTRIBUTION: Congo (DR).

ciliata Delève, 1965b

Helminthopsis ciliata Delève, 1965b: 12

TYPE LOCALITY: South Africa.

DISTRIBUTION: South Africa.

compacta Delève, 1965b

Helminthopsis compacta Delève, 1965b: 25 (subgenus *Elmidoliana*)

TYPE LOCALITY: Congo (DR).

DISTRIBUTION: Congo (DR).

dissimilis Grouvelle, 1906c

Helminthopsis dissimilis Grouvelle, 1906c: 320

Type Locality: Kenya.

Distribution: Kenya.

elegans Alluaud, 1933

Helminthopsis elegans Alluaud, 1933: 158

Type Locality: Burkina Faso.

Distribution: Angola, Burkina Faso, Congo (DR), Ghana, Ivory Coast, Liberia.

elongata Delève, 1965b

Helminthopsis elongata Delève, 1965b: 23 (subgenus *Elmidoliana*)

Type Locality: Congo (DR).

Distribution: Congo (DR), South Africa, Zambia.

fallaciosa Delève, 1965b

Helminthopsis fallaciosa Delève, 1965b: 8

Type Locality: Guinea.

Distribution: Guinea.

Note: *Helminthopsis fallaciosa* is possibly a junior synonym of *Helminthopsis dissimilis* Grouvelle, 1906c (see Delève 1965b: 21).

gracilis Delève, 1945b

Helminthopsis gracilis Delève, 1945b: 8

Type Locality: Congo (DR).

Distribution: Angola, Congo (DR), Congo (R), Ghana, Ivory Coast.

hypocrita Delève, 1965b

Helminthopsis hypocrita Delève, 1965b: 14

Type Locality: Congo (DR).

Distribution: Congo (DR).

inornata Delève, 1968b

Helminthopsis inornata Delève, 1968b: 203 (subgenus *Elmidoliana*)

Type Locality: Ivory Coast.

Distribution: Ivory Coast.

interposita Delève, 1965b

Helminthopsis interposita Delève, 1965b: 21

Type Locality: Congo (DR).

Distribution: Congo (DR).

lepida Delève, 1965b

Helminthopsis lepida Delève, 1965b: 6

Type Locality: Guinea.

Distribution: Guinea.

lucida Grouvelle, 1906c

Helminthopsis lucida Grouvelle, 1906c: 319

Type Locality: Kenya.

Distribution: Angola, Congo (DR), Congo (R), Ghana, Guinea, Ivory Coast, Kenya, South Africa; erroneously recorded from Burkina Faso by Alluaud (1933: 159).

luteopicta luteopicta Delève, 1938

Helminthopsis luteopicta Delève, 1938: 364 (subgenus *Elmidoliana*)

Type Locality: Congo (DR).

Distribution: Cameroon, Congo (DR), Congo (R).

luteopicta nigeriana Delève, 1974

Helminthopsis luteopicta nigeriana Delève, 1974: 282 (subgenus *Elmidoliana*)

Type Locality: Nigeria.

Distribution: Nigeria.

machadoi Delève, 1966d

Helminthopsis machadoi Delève, 1966d: 58 (subgenus *Elmidoliana*)

Type Locality: Angola.

Distribution: Angola.

marginalis Delève, 1973b

Helminthopsis marginalis Delève, 1973b: 309

Type Locality: Liberia.

Distribution: Liberia.

medleri Delève, 1974

Helminthopsis medleri Delève, 1974: 279 (subgenus *Elmidoliana*)

Type Locality: Nigeria.

Distribution: Nigeria.

melanaria Delève, 1967c

Helminthopsis melanaria Delève, 1967c: 75

Type Locality: Congo (R).

Distribution: Congo (R).

micros Delève, 1974

Helminthopsis micros Delève, 1974: 278 (subgenus *Elmidoliana*)

Type Locality: Ghana.

Distribution: Ghana.

molesta (Grouvelle, 1920)

Microdinodes molesta Grouvelle, 1920: 211 (subgenus *Elmidoliana*)

Type Locality: Tanzania.

Distribution: Angola, Tanzania.

perfida Delève, 1967a

Helminthopsis perfida Delève, 1967a: 335

Type Locality: Congo (R).

Distribution: Congo (R).

perplexa Delève, 1945b

Helminthopsis perplexa Delève, 1945b: 3

Type Locality: Congo (DR).

Distribution: Congo (DR), Congo (R), Gabon.

placita Delève, 1968b

Helminthopsis placita Delève, 1968b: 201

Type Locality: Ivory Coast.

Distribution: Ghana, Ivory Coast.

propinqua Delève, 1974

Helminthopsis propinqua Delève, 1974: 275 (subgenus *Elmidoliana*)

Type Locality: Ghana.

Distribution: Ghana.

proxima Delève, 1945b

Helminthopsis proxima Delève, 1945b: 10

Type Locality: Kenya.

Distribution: Kenya.

Note: Originally described in a key; formal description published by Delève (1946: 325).

punctulata Delève, 1945b

Helminthopsis punctulata Delève, 1945b: 7

Type Locality: Congo (DR).

DISTRIBUTION: Angola, Burkina Faso, Central African Republic, Congo (DR), Congo (R), Ghana, Ivory Coast, Liberia.

quadrinotata Delève, 1965b

Helminthopsis quadrinotata Delève, 1965b: 18

TYPE LOCALITY: Congo (R).

DISTRIBUTION: Congo (R), Liberia.

reticulata Delève, 1945b

Helminthopsis reticulata Delève, 1945b: 5

TYPE LOCALITY: Congo (DR).

DISTRIBUTION: Congo (DR), Congo (R), Ghana, Liberia.

rhodesiana Delève, 1965b

Helminthopsis rhodesiana Delève, 1965b: 29 (subgenus *Elmidoliana*)

TYPE LOCALITY: Zimbabwe.

DISTRIBUTION: Zimbabwe.

subglobosa Delève, 1967a

Helminthopsis subglobosa Delève, 1967a: 336

TYPE LOCALITY: Congo (R).

DISTRIBUTION: Congo (R).

zambezica arcuata Delève, 1974

Helminthopsis zambezica arcuata Delève, 1974: 280 (subgenus *Elmidoliana*)

TYPE LOCALITY: Ghana.

DISTRIBUTION: Ghana.

zambezica zambezica Delève, 1965b

Helminthopsis zambezica Delève, 1965b: 27 (subgenus *Elmidoliana*)

TYPE LOCALITY: Zambia.

DISTRIBUTION: Angola, Cameroon, Congo (R), Gabon, Ivory Coast, Liberia, Zambia.

Heterelmis Sharp, 1882

Heterelmis Sharp, 1882: 130. — Type species: *Heterelmis obscura* Sharp, 1882. — Gender feminine.

NOTE: Incorrect subsequent spelling: "*Heterhelmis*" Zaitzev (1908: 309), not to be regarded as an emendation (ICZN 1999: Art. 33.2.1).

apicata (Grouvelle, 1896a)

Helmis apicata Grouvelle, 1896a: 49

Type Locality: Ecuador.

Distribution: Ecuador.

comalensis Bosse, Tuff & Brown, 1988

Heterelmis comalensis Bosse, Tuff & Brown, 1988: 199

Type Locality: USA (Texas).

Distribution: USA (Texas).

convexicollis Delève, 1968a

Heterelmis convexicollis Delève, 1968a: 249

Type Locality: Ecuador.

Distribution: Colombia, Ecuador.

dubia Grouvelle, 1889b

Heterelmis dubia Grouvelle, 1889b: 399

Type Locality: Brazil (Santa Catarina).

Distribution: Brazil (Santa Catarina).

gibbosa (Grouvelle, 1889b) **comb.n.**

Helmis gibbosa Grouvelle, 1889b: 404

Type Locality: Brazil (Rio de Janeiro).

Distribution: Brazil (Rio de Janeiro).

Note: This species is transferred herewith from *Elmis* Latreille, 1802.

glabra (Horn, 1870)

Elmis glabra Horn, 1870: 37

Type Locality: USA (Arizona).

Distribution: Belize, Costa Rica, Mexico, Nicaragua (Manzo 2013: 211), USA (Arizona, Texas).

Note: Incorrect original spelling: "*Elmis glaber*".

Synonym: *Heterelmis acicula* Hinton, 1940a: 389. — Type locality: Mexico.

impressicollis Delève, 1968a

Heterelmis impressicollis Delève, 1968a: 252

Type Locality: Ecuador.

Distribution: Ecuador.

limnoides Hinton, 1936b

Heterelmis limnoides Hinton, 1936b: 288

TYPE LOCALITY: Brazil (Santa Catarina).

DISTRIBUTION: Brazil (Santa Catarina).

longula Sharp, 1887

Heterelmis longula Sharp, 1887: 775

TYPE LOCALITY: Mexico.

DISTRIBUTION: Mexico.

NOTE: Incorrect original spelling: "*longulus*".

lucida Delève, 1968a

Heterelmis lucida Delève, 1968a: 248

TYPE LOCALITY: Ecuador.

DISTRIBUTION: Ecuador.

neglecta Grouvelle, 1896a

Heterelmis neglecta Grouvelle, 1896a: 46

TYPE LOCALITY: Bolivia.

DISTRIBUTION: Bolivia, Ecuador.

obesa obesa Sharp, 1882

Heterelmis obesa Sharp, 1882: 131

TYPE LOCALITY: Guatemala.

DISTRIBUTION: Costa Rica, Guatemala, Mexico, Nicaragua, Peru, USA (Arizona, California, New Mexico, Texas).

NOTE: Incorrect original spelling: "*obesus*".

obesa plana Hinton, 1936b

Heterelmis obscura [!] *plana* Hinton, 1936b: 289

TYPE LOCALITY: Mexico.

DISTRIBUTION: Mexico.

NOTE: Originally described as "*H. obscura* Sharp var. *plana* Hinton (MSS.)", listed under "*Heterelmis obesa* var. *plana* Hinton (1937)" by Hinton (1937a: 111); formal description published by Hinton (1940a: 385) under the name "*Heterelmis obesa plana*, subsp. n."; probably a discrete species.

obscura Sharp, 1882

Heterelmis obscura Sharp, 1882: 130

TYPE LOCALITY: Guatemala.

DISTRIBUTION: Brazil (Santa Catarina), Costa Rica, Guatemala, Mexico, Peru.

NOTE: Incorrect original spelling: "*obscurus*".

pusilla Delève, 1968a

Heterelmis pusilla Delève, 1968a: 255

Type Locality: Ecuador.

Distribution: Ecuador.

simplex codrus Hinton, 1971d

Heterelmis simplex codrus Hinton, 1971d: 263

Type Locality: Trinidad and Tobago (Trinidad).

Distribution: Trinidad and Tobago (Tobago, Trinidad).

Note: Epithet name is a noun in apposition (king of Athens in Greek mythology).

simplex simplex Sharp, 1882

Heterelmis simplex Sharp, 1882: 131

Type Locality: Guatemala.

Distribution: Costa Rica, Guatemala, Peru.

stephani Brown, 1972b

Heterelmis stephani Brown, 1972b: 230

Type Locality: USA (Arizona).

Distribution: USA (Arizona).

tarsalis Hinton, 1940a

Heterelmis tarsalis Hinton, 1940a: 374

Type Locality: Mexico.

Distribution: Mexico.

trivialis (Germain, 1892)

Elmis trivialis Germain, 1892: 245

Type Locality: Chile.

Distribution: Chile.

vilcanota Spangler, 1980b

Heterelmis vilcanota Spangler, 1980b: 205

Type Locality: Peru.

Distribution: Peru.

vulnerata (Leconte, 1874)

Elmis vulnerata Leconte, 1874: 53

Type Locality: USA (Texas).

Distribution: Mexico, USA (Kansas, Oklahoma, Texas).

Note: Incorrect original spelling: "*Elmis vulneratus*".

Heterlimnius Hinton, 1935

Heterlimnius Hinton, 1935: 178. — Type species: *Helmis koebelei* Martin, 1927. — Gender masculine.

SYNONYM: *Cyclolimnius* Nomura, 1958b: 54. — Type species: *Optioservus kubotai* Nomura, 1958b. — Gender masculine. NOTE: Originally described as subgenus of *Optioservus*.

amabilis Kamite, 2011

Heterlimnius amabilis Kamite, 2011: 409

TYPE LOCALITY: China (Sichuan).

DISTRIBUTION: China (Guizhou, Sichuan).

ater (Nomura, 1958b)

Optioservus ater Nomura, 1958b: 56

TYPE LOCALITY: Japan (Honshu).

DISTRIBUTION: Japan (Honshu).

SYNONYM: *Optioservus hayashii* Nomura, 1960: 34. — Type locality: Japan (Honshu). NOTE: Synonymized by Kamite (2009: 212).

corpulentus (Leconte, 1874)

Elmis corpulenta Leconte, 1874: 52

TYPE LOCALITY: Canada (British Columbia).

DISTRIBUTION: Canada (Alberta, British Columbia), USA (California, Colorado, Idaho, Montana, Nevada, New Mexico, Oregon, South Dakota, Utah, Washington, Wyoming).

NOTE: Incorrect original spelling: "*Elmis corpulentus*"; incorrect subsequent spelling: "*Heterlimnius corpulenta*" (Poole & Gentili 1996: 264).

SYNONYMS: *Elmis antennata* Fall, 1907: 227. — Type locality: USA (New Mexico). NOTE: Incorrect original spelling: "*Elmis antennatus*".

Helmis koebelei Martin, 1927: 68. — Type locality: USA (Washington). NOTE. Incorrect subsequent spelling: "*koebeli*" (Barr 2011: 206, 208).

ennearthrus Kamite, 2009

Heterlimnius ennearthrus Kamite, 2009: 216

TYPE LOCALITY: Kazakhstan.

DISTRIBUTION: Kazakhstan, Russia.

hasegawai (Nomura, 1958b)

Optioservus hasegawai Nomura, 1958b: 52

TYPE LOCALITY: Russia (Sakhalin).

DISTRIBUTION: China (Jilin, Liaoning), Japan (Hokkaido, Honshu), (Korea (DPR)), Korea (R), Russia (Far East, incl. Sakhalin and Kuril Islands).

SYNONYMS: *Optioservus kubotai* Nomura, 1958b: 54. — Type locality: Japan (Honshu).

Optioservus kubotai saghaliensis Nomura, 1958b: 55. — Type locality: Russia (Sakhalin).

hisamatsui Kamite, 2009

Heterlimnius hisamatsui Kamite, 2009: 219

Type Locality: China (Sichuan).

Distribution: China (Sichuan, Yunnan).

horii Kamite, 2012

Heterlimnius horii Kamite, 2012: 291

Type Locality: China (Sichuan).

Distribution: China (Sichuan).

ikedai Kamite, 2011

Heterlimnius ikedai Kamite, 2011: 411

Type Locality: China (Guizhou).

Distribution: China (Guizhou).

jaechi Kamite, 2009

Heterlimnius jaechi Kamite, 2009: 218

Type Locality: India (Himachal Pradesh).

Distribution: Bhutan, India (Himachal Pradesh).

quadrigibbus Kamite, 2012

Heterlimnius quadrigibbus Kamite, 2012: 292

Type Locality: China (Tibet).

Distribution: China (Tibet).

shepardi Kamite, 2009

Heterlimnius shepardi Kamite, 2009: 221

Type Locality: China (Shaanxi).

Distribution: China (Guizhou, Shaanxi, Sichuan).

trachys (Janssens, 1959a)

Stenelmis trachys Janssens, 1959a: 5

Type Locality: Afghanistan.

Distribution: Afghanistan.

vietnamensis Kamite, 2011

Heterlimnius vietnamensis Kamite, 2011: 412

Type Locality: Vietnam.

Distribution: Vietnam.

yokoii Kamite, 2012

Heterlimnius yokoii Kamite, 2012: 294

TYPE LOCALITY: China (Yunnan).

DISTRIBUTION: China (Yunnan).

Hexacylloepus Hinton, 1940a

Hexacylloepus Hinton, 1940a: 331. — Type species: *Helmis smithi* Grouvelle, 1898a. — Gender masculine.

abditus (Hinton, 1937a)

Cylloepus abditus Hinton, 1937a: 106

TYPE LOCALITY: Mexico.

DISTRIBUTION: Mexico, USA (Arizona).

abdominalis (Hinton, 1937d)

Cylloepus abdominalis Hinton, 1937d: 284

TYPE LOCALITY: Brazil (Santa Catarina).

DISTRIBUTION: Brazil (Santa Catarina).

aciculus (Hinton, 1937d)

Cylloepus aciculus Hinton, 1937d: 282

TYPE LOCALITY: Brazil (Santa Catarina).

DISTRIBUTION: Brazil (Santa Catarina), Paraguay.

apicalis Hinton, 1940a

Hexacylloepus apicalis Hinton, 1940a: 334

TYPE LOCALITY: Mexico.

DISTRIBUTION: Mexico.

bassindalei Hinton, 1969

Hexacylloepus bassindalei Hinton, 1969: 127

TYPE LOCALITY: Brazil (Santa Catarina).

DISTRIBUTION: Brazil (Santa Catarina), Paraguay.

danforthi (Musgrave, 1935) **comb.n.**

Cylloepus danforthi Musgrave, 1935: 33

TYPE LOCALITY: Puerto Rico.

DISTRIBUTION: Puerto Rico.

NOTE: This species is transferred herewith from *Cylloepus* Erichson, 1847.

ferrugineus (Horn, 1870)

Elmis ferruginea Horn, 1870: 39

Type Locality: USA (Texas).

Distribution: Belize, USA (New Mexico, Oklahoma, Texas).

Note: Incorrect original spelling: "*Elmis ferrugineus*"; incorrect subsequent spelling: "*Hexacylloepus ferruginea*" (Poole & Gentili 1996: 264).

filiformis (Darlington, 1927)

Helmis filiformis Darlington, 1927: 94

Type Locality: Cuba.

Distribution: Cuba, Jamaica.

Synonym: *Helmis filiformis jamaicensis* Darlington, 1927: 97. — Type locality: Jamaica. Note: Regarded as a possible synonym of *Hexacylloepus filiformis* (Darlington, 1927) by Darlington (1936: 79).

flavipes (Grouvelle, 1889b)

Helmis flavipes Grouvelle, 1889b: 404

Type Locality: Brazil (Santa Catarina).

Distribution: Brazil (Santa Catarina).

frater Hinton, 1939c

Hexacylloepus frater Hinton, 1939c: 181

Type Locality: Brazil (Santa Catarina).

Distribution: Brazil (Santa Catarina), French Guiana, Paraguay.

granosus (Grouvelle, 1889b)

Helmis granosus Grouvelle, 1889b: 403

Type Locality: Brazil (Santa Catarina).

Distribution: Brazil (Santa Catarina).

granulosus (Sharp, 1882)

Elmis granulosa Sharp, 1882: 136

Type Locality: Guatemala.

Distribution: Costa Rica, Guatemala, Panama.

Note: Incorrect original spelling: "*Elmis granulosus*".

haitianus (Darlington, 1936) **comb.n.**

Helmis haitiana Darlington, 1936: 80

Type Locality: Haiti.

Distribution: Haiti.

Note: This species is transferred herewith from *Cylloepus* Erichson, 1847.

heterelmoides Hinton, 1939c

Hexacylloepus heterelmoides Hinton, 1939c: 183

Type Locality: French Guiana.

Distribution: French Guiana.

horni (Hinton, 1937a)

Cylloepus horni Hinton, 1937a: 109

Type Locality: Mexico.

Distribution: Mexico, Nicaragua.

indistinctus (Hinton, 1937d)

Cylloepus indistinctus Hinton, 1937d: 282

Type Locality: Brazil (Santa Catarina).

Distribution: Brazil (Santa Catarina), Paraguay.

lahottensis (Darlington, 1936) **comb.n.**

Helmis lahottensis Darlington, 1936: 81

Type Locality: Haiti.

Distribution: Haiti.

Note: This species is transferred herewith from *Cylloepus* Erichson, 1847.

nirgua Hinton, 1973b

Hexacylloepus nirgua Hinton, 1973b: 252

Type Locality: Venezuela.

Distribution: Ecuador, Venezuela.

nothrus Spangler, 1966

Hexacylloepus nothrus Spangler, 1966: 408

Type Locality: Peru.

Distribution: Bolivia, Peru.

nunezi Hinton, 1973b

Hexacylloepus nunezi Hinton, 1973b: 253

Type Locality: Venezuela.

Distribution: Venezuela.

plaumanni (Hinton, 1937d)

Cylloepus plaumanni Hinton, 1937d: 280

Type Locality: Brazil (Santa Catarina).

Distribution: Brazil (Santa Catarina).

quadratus (Darlington, 1927)

Helmis quadrata Darlington, 1927: 95

Type Locality: Cuba.

Distribution: Cuba.

scabrosus Hinton, 1940a

Hexacylloepus scabrosus Hinton, 1940a: 337

Type Locality: Mexico.

Distribution: Mexico.

smithi bejuma Hinton, 1973b

Hexacylloepus smithi bejuma Hinton, 1973b: 255

Type Locality: Venezuela.

Distribution: Venezuela.

smithi smithi (Grouvelle, 1898a)

Helmis smithi Grouvelle, 1898a: 47

Type Locality: Grenada.

Distribution: Grenada, Trinidad and Tobago (Tobago, Trinidad).

subsulcatus (Grouvelle, 1889b)

Helmis subsulcata Grouvelle, 1889b: 403

Type Locality: Brazil (Santa Catarina).

Distribution: Brazil (Santa Catarina).

sulcatus (Grouvelle, 1889b)

Helmis sulcata Grouvelle, 1889b: 402

Type Locality: Brazil (Santa Catarina).

Distribution: Brazil (Santa Catarina).

Hexanchorus Sharp, 1882

Hexanchorus Sharp, 1882: 127. — Type species: *Hexanchorus gracilipes* Sharp, 1882. — Gender masculine.

Note: Incorrect subsequent spellings: "*Xexanchorus*" (Grouvelle 1896c: 78), "*Xexanchrous*" (Spangler & Staines 2004b: 45).

angeli Laššová, Čiampor & Čiamporová-Zaťovičová, 2014

Hexanchorus angeli Laššová, Čiampor & Čiamporová-Zaťovičová, 2014: 188

Type Locality: Venezuela.

Distribution: Venezuela.

bifurcatus Maier & Short, 2014

Hexanchorus bifurcatus Maier & Short, 2014: 138

Type Locality: Suriname.

Distribution: Suriname.

browni Spangler & Santiago-Fragoso, 1992

Hexanchorus browni Spangler & Santiago-Fragoso, 1992: 39

Type Locality: Panama.

Distribution: Costa Rica, Mexico, Panama.

caraibus (Coquerel, 1851)

Potamophilus caraibus Coquerel, 1851: 601

Type Locality: "Antilles".

Distribution: Brazil (Rio de Janeiro), Dominica, Guadeloupe, Martinque, Saint Lucia, Saint Vincent.

cordillierae (Guérin Méneville, 1843)

Potamophilus cordillierae Guérin Méneville, 1843: 19

Type Locality: Colombia.

Distribution: Colombia.

Note: Incorrect original spelling: "*cordillieræ*"; incorrect subsequent spellings: "*cordilieræ*" (Lacordaire 1854: 502), "*cordillerae*" (Spangler & Staines 2004b: 45).

crinitus Spangler & Santiago-Fragoso, 1992

Hexanchorus crinitus Spangler & Santiago-Fragoso, 1992: 31

Type Locality: Costa Rica.

Distribution: Costa Rica, Panama.

dentitibialis Maier, 2013

Hexanchorus dentitibialis Maier, 2013: 43

Type Locality: Venezuela.

Distribution: Venezuela.

dimorphus Spangler & Staines, 2004b

Hexanchorus dimorphus Spangler & Staines, 2004b: 46

Type Locality: Argentina.

Distribution: Argentina, Paraguay.

emarginatus Spangler & Santiago-Fragoso, 1992

Hexanchorus emarginatus Spangler & Santiago-Fragoso, 1992: 37

Type Locality: Costa Rica.

Distribution: Costa Rica, Panama.

falconensis Maier, 2013

Hexanchorus falconensis Maier, 2013: 46

Type Locality: Venezuela.

Distribution: Venezuela.

flintorum Maier, 2013

Hexanchorus flintorum Maier, 2013: 49

Type Locality: Venezuela.

Distribution: Venezuela.

gracilipes Sharp, 1882

Hexanchorus gracilipes Sharp, 1882: 128

Type Locality: Mexico.

Distribution: Brazil (Rio de Janeiro) (Passos et al. 2010b: 380), Costa Rica, Guatemala, Mexico, Panama.

Synonym: *Hexanchorus gracilipes orientalis* Zaragoza Caballero, 1982: 354. — Type locality: Mexico.

homaeotarsoides Maier, 2013

Hexanchorus homaeotarsoides Maier, 2013: 52

Type Locality: Venezuela.

Distribution: Venezuela.

inflatus Maier, 2013

Hexanchorus inflatus Maier, 2013: 55

Type Locality: Venezuela.

Distribution: Venezuela.

leleupi Delève, 1968a

Hexanchorus leleupi Delève, 1968a: 214

Type Locality: Ecuador.

Distribution: Ecuador.

mcdiarmidi Spangler & Staines, 2004b

Hexanchorus mcdiarmidi Spangler & Staines, 2004b: 47

Type Locality: Venezuela.

Distribution: Venezuela.

Note: Incorrect subsequent spelling: "*mediarmidi*" (Segura et al. 2013: 47); this species was named after Roy McDiarmid.

shannoni Spangler & Staines, 2004b

Hexanchorus shannoni Spangler & Staines, 2004b: 45

TYPE LOCALITY: Argentina.

DISTRIBUTION: Argentina.

tarsalis Hinton, 1937e

Hexanchorus tarsalis Hinton, 1937e: 95

TYPE LOCALITY: Brazil (Santa Catarina).

DISTRIBUTION: Brazil (Santa Catarina).

thermarius (Coquerel, 1851)

Potamophilus thermarius Coquerel, 1851: 604

TYPE LOCALITY: Brazil (details unknown).

DISTRIBUTION: Brazil (Espírito Santo).

tibialis Hinton, 1935

Hexanchorus tibialis Hinton, 1935: 176

TYPE LOCALITY: Bolivia.

DISTRIBUTION: Bolivia, Peru.

usitatus Spangler & Santiago-Fragoso, 1992

Hexanchorus usitatus Spangler & Santiago-Fragoso, 1992: 33

TYPE LOCALITY: Panama.

DISTRIBUTION: Costa Rica, Nicaragua, Panama.

Hintonelmis Spangler, 1966

Hintonelmis Spangler, 1966: 411. — Type species: *Hintonelmis sandersoni* Spangler, 1966. — Gender feminine.

anamariae Fernandes, Passos & Hamada, 2010a

Hintonelmis anamariae Fernandes, Passos & Hamada, 2010a: 44

TYPE LOCALITY: Brazil (Amazonas).

DISTRIBUTION: Brazil (Amazonas).

atys Hinton, 1971a

Hintonelmis atys Hinton, 1971a: 201

TYPE LOCALITY: Brazil (Amazonas).

DISTRIBUTION: Brazil (Amazonas); erroneously recorded from the Brazilian State of Rondônia by Hinton (1971a: 202) and Segura et al. (2013: 23).

carus Hinton, 1971a

Hintonelmis carus Hinton, 1971a: 203

Type Locality: Brazil (Amazonas).

Distribution: Brazil (Amazonas, Rondônia); erroneously recorded from the Brazilian State of Mato Grosso by Hinton (1971a: 204) and Segura et al. (2013: 23).

Note: The epithet name is a noun in apposition (Roman Emperor).

delevei Hinton, 1971a

Hintonelmis delevei Hinton, 1971a: 205

Type Locality: French Guiana.

Distribution: Brazil (Amazonas), French Guiana.

maro Hinton, 1971a

Hintonelmis maro Hinton, 1971a: 204

Type Locality: Brazil (Pará).

Distribution: Brazil (Amazonas, Pará).

messa Hinton, 1971a

Hintonelmis messa Hinton, 1971a: 208

Type Locality: Brazil (Pará).

Distribution: Brazil (Pará), Paraguay.

opis Hinton, 1971a

Hintonelmis opis Hinton, 1971a: 199

Type Locality: Brazil (Amazonas).

Distribution: Brazil (Amazonas, Rondônia); erroneously recorded from the Brazilian State of Mato Grosso by Hinton (1971a: 200).

perfecta (Grouvelle, 1908)

Ancyronyx perfectus Grouvelle, 1908: 185

Type Locality: French Guiana.

Distribution: French Guiana.

Note: Incorrect subsequent spelling: "*Hintonelmis perfectus*" (Hinton 1971a: 202); this species was transferred to *Hintonelmis* by Delève (1970c: 65).

sandersoni Spangler, 1966

Hintonelmis sandersoni Spangler, 1966: 412

Type Locality: Peru.

Distribution: Peru.

sloanei Hinton, 1971a

Hintonelmis sloanei Hinton, 1971a: 198

Type Locality: Brazil (Pará).

Distribution: Brazil (Amazonas, Pará).

sul Hinton, 1971a

Hintonelmis sul Hinton, 1971a: 196

Type Locality: Brazil (Rondônia); erroneously, Hinton (1971a: 197) attributed the type locality to Mato Grosso ("Mato Grosso, Guyara-Mirim, viii.1937 (H.E. Hinton)").

Distribution: Brazil (Rondônia); erroneously reported from Mato Grosso by Segura et al. (2013: 23); this species is known only from the type locality.

Hispaniolara Brown, 1981c

Hispaniolara Brown, 1981c: 85. — Type species: *Hispaniolara farri* Brown, 1981c. — Gender feminine.

farri Brown, 1981c

Hispaniolara farri Brown, 1981c: 88

Type Locality: Dominican Republic.

Distribution: Dominican Republic.

Holcelmis Hinton, 1973a

Holcelmis Hinton, 1973a: 1. — Type species: *Holcelmis woodruffi* Hinton, 1973a. — Gender feminine.

mamore Hinton, 1973a

Holcelmis mamore Hinton, 1973a: 4

Type Locality: Bolivia.

Distribution: Bolivia.

woodruffi Hinton, 1973a

Holcelmis woodruffi Hinton, 1973a: 3

Type Locality: Bolivia.

Distribution: Bolivia.

Homalosolus Jäch & Kodada, 1996b

Homalosolus Jäch & Kodada, 1996b: 400. — Type species: *Homalosolus hospitalis* Jäch & Kodada, 1996b. — Gender masculine.

felis Jäch & Kodada, 1996b

Homalosolus felis Jäch & Kodada, 1996b: 405

Type Locality: Malaysia (Sarawak).

Distribution: Malaysia (Sarawak).

heissi Jäch & Kodada, 1996b

Homalosolus heissi Jäch & Kodada, 1996b: 405

Type Locality: Brunei.

Distribution: Brunei.

hospitalis Jäch & Kodada, 1996b

Homalosolus hospitalis Jäch & Kodada, 1996b: 403

Type Locality: Malaysia (Sarawak).

Distribution: Malaysia (Sarawak).

zitnanskae Čiampor, 1999

Homalosolus zitnanskae Čiampor, 1999: 31

Type Locality: Malaysia (Sabah).

Distribution: Malaysia (Sabah).

Huleechius Brown, 1981b

Huleechius Brown, 1981b: 230. — Type species: *Cylloepus spinipes* Hinton, 1934. — Gender masculine.

marroni carolus Brown, 1981b

Huleechius marroni carolus Brown, 1981b: 239

Type Locality: USA (Arizona).

Distribution: USA (Arizona).

Note: Might be a discrete species; synonymized with *H. marroni* by Poole & Gentili (1996: 264) without presenting any evidence.

marroni marroni Brown, 1981b

Huleechius marroni Brown, 1981b: 237

Type Locality: Mexico.

Distribution: Mexico.

spinipes (Hinton, 1934)

Cylloepus spinipes Hinton, 1934: 192

Type Locality: Mexico.

Distribution: Mexico.

Hydora Anon. [Broun], 1882

Hydora Anon. [Broun], 1882: 409. — Type species: *Pachycephala picea* Broun, 1881. — Gender feminine.

NOTE: The name *Hydora* is a substitute name for *Pachycephala* Broun, 1881; originally published anonymously (see Anonymous [Broun] 1882: 409), no author is named, neither in the article itself, nor in the table of contents of the particular volume; in the index to volume IX (pp. 480–484) *Hydora* is not listed at all; it can, however, be assumed, that Broun himself suggested this name (see ICZN 1999: Recommendation 51D for citation of anonymous authorship); anonymous authorship before 1951 does not prevent availability (ICZN 1999: Art. 14).

HOMONYM: *Pachycephala* Broun, 1881: 672. — Type species: *Pachycephala picea* Broun, 1881. — Gender feminine. NOTE: Junior homonym of *Pachycephala* Vigors, 1825: 444 (Aves).

SYNONYM: *Udorus* Broun, 1882: 128. — Type species: *Pachycephala picea* Broun, 1881. — Gender masculine. NOTE: Substitute name for *Pachycephala* Broun, 1881; incorrect subsequent spelling: "*Udorius*" (Zaitzev 1908: 289); the substitute names *Hydora* and *Udorus* were published simultaneously in May 1882; however, Broun (1885) used only the name *Hydora* and must therefore be regarded as "First Reviser" (ICZN 1999: Art. 24.2).

angusticollis (Pascoe, 1877)

Pomatinus angusticollis Pascoe, 1877: 141 ("*Potaminus*" [Dryopidae])

TYPE LOCALITY: New Zealand.

DISTRIBUTION: New Zealand.

SYNONYM: *Hydora vestita* Broun, 1914: 153. — Type locality: New Zealand. NOTE: Synonymized by Hinton (1935: 174).

annectens Spangler & Brown, 1981

Hydora annectens Spangler & Brown, 1981: 599

TYPE LOCALITY: Chile.

DISTRIBUTION: Argentina, Chile.

lanigera Broun, 1914

Hydora lanigera Broun, 1914: 154

TYPE LOCALITY: New Zealand.

DISTRIBUTION: New Zealand.

laticeps (Carter & Zeck, 1932)

Stetholus laticeps Carter & Zeck, 1932: 202

TYPE LOCALITY: Australia (New South Wales).

DISTRIBUTION: Australia (New South Wales).

NOTE: This species probably belongs to an undescribed genus.

lenta Spangler & Brown, 1981

Hydora lenta Spangler & Brown, 1981: 602

TYPE LOCALITY: Chile.

DISTRIBUTION: Chile.

nitida Broun, 1885

Hydora nitida Broun, 1885: 385

TYPE LOCALITY: New Zealand.

DISTRIBUTION: New Zealand.

obsoleta Broun, 1885

Hydora obsoleta Broun, 1885: 385

TYPE LOCALITY: New Zealand.

DISTRIBUTION: New Zealand.

picea (Broun, 1881)

Pachycephala picea Broun, 1881: 672

TYPE LOCALITY: New Zealand.

DISTRIBUTION: New Zealand.

NOTE: Incorrect original spelling: "*Pachycephala piceum*".

subaenea Broun, 1914

Hydora subaenea Broun, 1914: 154

TYPE LOCALITY: New Zealand.

DISTRIBUTION: New Zealand.

Hydrethus Fairmaire, 1889b

Hydrethus Fairmaire, 1889b: XC[90]. — Type species: *Hydrethus dermestoides* Fairmaire, 1889b. — Gender masculine.

dermestoides Fairmaire, 1889b

Hydrethus dermestoides Fairmaire, 1889b: XC[90]

TYPE LOCALITY: Madagascar.

DISTRIBUTION: Madagascar.

elouardi Bameul, 1996

Hydrethus elouardi Bameul, 1996: 279

TYPE LOCALITY: Madagascar.

DISTRIBUTION: Madagascar.

perrieri Delève, 1963c

Hydrethus perrieri Delève, 1963c: 449

Type Locality: Madagascar.

Distribution: Madagascar.

proximus Delève, 1963c

Hydrethus proximus Delève, 1963c: 451

Type Locality: Madagascar.

Distribution: Madagascar.

Hypsilara Maier & Spangler, 2011

Hypsilara Maier & Spangler, 2011: 27. — Type species: *Hypsilara royi* Maier & Spangler, 2011. — Gender feminine.

autanai Laššová, Čiampor & Čiamporová-Zaťovičová, 2014

Hypsilara autanai Laššová, Čiampor & Čiamporová-Zaťovičová, 2014: 192

Type Locality: Venezuela.

Distribution: Venezuela.

breweri Čiampor, Laššová & Čiamporová-Zaťovičová, 2013

Hypsilara breweri Čiampor, Laššová & Čiamporová-Zaťovičová, 2013: 593

Type Locality: Venezuela.

Distribution: Venezuela.

royi Maier & Spangler, 2011

Hypsilara royi Maier & Spangler, 2011: 33

Type Locality: Venezuela.

Distribution: Venezuela.

Ilamelmis Delève, 1973a

Ilamelmis Delève, 1973a: 14. — Type species: *Ilamelmis brunnescens* Delève, 1973a. — Gender feminine.

Note: This genus is possibly a junior synonym of *Elmidolia* Fairmaire, 1897b.

Synonym: *Aruelmis* Delève, 1973d: 71. — Type species: *Aruelmis starmuhlneri* Delève, 1973d. – Gender feminine.

brunnescens Delève, 1973a

Ilamelmis brunnescens Delève, 1973a: 14

TYPE LOCALITY: Sri Lanka.
DISTRIBUTION: Sri Lanka.

crassa Delève, 1973a

Ilamelmis crassa Delève, 1973a: 16
TYPE LOCALITY: Sri Lanka.
DISTRIBUTION: Sri Lanka.

foveicollis (Grouvelle, 1896d)

Helmis foveicollis Grouvelle, 1896d: 45
TYPE LOCALITY: Sri Lanka.
DISTRIBUTION: Sri Lanka.

starmuhlneri (Delève, 1973d)

Aruelmis starmuhlneri Delève, 1973d: 72
TYPE LOCALITY: Sri Lanka.
DISTRIBUTION: Sri Lanka.
NOTE: Incorrect original spellings: "*starmuhlneri*" (5 ×) and "*stamuhlneri*" (1 ×).

Indosolus Bollow, 1940b

Indosolus Bollow, 1940b: 28. — Type species: *Esolus nitidus* Bollow, 1940b. — Gender masculine.
NOTE: Originally described as subgenus of *Esolus*, elevated to generic rank by Jäch & Boukal (1995a: 308).

acutangulus (Champion, 1923)

Zaitzevia acutangula Champion, 1923: 171
TYPE LOCALITY: India (Uttarakhand).
DISTRIBUTION: India (Uttarakhand).

nitidus (Bollow, 1940b)

Esolus nitidus Bollow, 1940b: 28
TYPE LOCALITY: Myanmar.
DISTRIBUTION: China (Yunnan), Myanmar.

Jaechomorphus Kodada, 1993b

Jaechomorphus Kodada, 1993b: 1. — Type species: *Jaechomorphus gracilis* Kodada, 1993b. — Gender masculine.

gigas Kodada, 1993b

Jaechomorphus gigas Kodada, 1993b: 6

TYPE LOCALITY: Indonesia (Sumatra).

DISTRIBUTION: Indonesia (Sumatra).

gracilis Kodada, 1993b

Jaechomorphus gracilis Kodada, 1993b: 4

TYPE LOCALITY: Indonesia (Sumatra).

DISTRIBUTION: Indonesia (Sumatra).

satyr Kodada, 1993b

Jaechomorphus satyr Kodada, 1993b: 5

TYPE LOCALITY: Indonesia (Sumatra).

DISTRIBUTION: Indonesia (Sumatra).

Jilanzhunychus Jäch & Boukal, 1995a

Jilanzhunychus Jäch & Boukal, 1995a: 304. — Type species: *Jilanzhunychus costatus* Jäch & Boukal, 1995a. — Gender masculine.

costatus Jäch & Boukal, 1995a

Jilanzhunychus costatus Jäch & Boukal, 1995a: 306

TYPE LOCALITY: China (Guangxi).

DISTRIBUTION: China (Guangxi).

Jolyelmis Spangler & Faitoute, 1991

Jolyelmis Spangler & Faitoute, 1991: 322. — Type species: *Jolyelmis auyana* Spangler & Faitoute, 1991. — Gender feminine.

auyana Spangler & Faitoute, 1991

Jolyelmis auyana Spangler & Faitoute, 1991: 325

TYPE LOCALITY: Venezuela.

DISTRIBUTION: Venezuela.

derkai Čiampor & Kodada, 1999

Jolyelmis derkai Čiampor & Kodada, 1999: 55

TYPE LOCALITY: Venezuela.

DISTRIBUTION: Venezuela.

reitmaieri Čiampor & Kodada, 1999

Jolyelmis reitmaieri Čiampor & Kodada, 1999: 57

Type Locality: Venezuela.

Distribution: Venezuela.

spangleri Kodada, Derka & Čiampor, 2012

Jolyelmis spangleri Kodada, Derka & Čiampor, 2012: 7

Type Locality: Venezuela.

Distribution: Venezuela.

Kingolus Carter & Zeck, 1929

Kingolus Carter & Zeck, 1929: 53. — Type species: *Elmis metallica* King, 1865. — Gender masculine.

aeratus (Carter, 1926b)

Helmis aerata Carter, 1926b: 62

Type Locality: Australia (Victoria).

Distribution: Australia (ACT, New South Wales, Victoria).

cupreus (Carter, 1926a)

Helmis cuprea Carter, 1926a: 507

Type Locality: Australia (New South Wales).

Distribution: Australia (New South Wales).

davisi Zeck, 1948

Kingolus davisi Zeck, 1948: 277

Type Locality: Australia (Queensland).

Distribution: Australia (Queensland).

flavoplagiatus Carter & Zeck, 1929

Kingolus flavoplagiatus Carter & Zeck, 1929: 54

Type Locality: Australia (New South Wales).

Distribution: Australia (ACT, New South Wales).

flavosignatus Carter & Zeck, 1929

Kingolus flavosignatus Carter & Zeck, 1929: 55

Type Locality: Australia (New South Wales).

Distribution: Australia (New South Wales).

heroni Carter & Zeck, 1929

Kingolus heroni Carter & Zeck, 1929: 56

TYPE LOCALITY: Australia (New South Wales).

DISTRIBUTION: Australia (ACT, New South Wales).

metallicus (King, 1865)

Elmis metallica King, 1865: 160

TYPE LOCALITY: Australia (New South Wales, South Australia, or Victoria); the type locality cannot be localized exactly ("Murray River" (King 1865: 160)).

DISTRIBUTION: Australia (ACT, New South Wales, South Australia?, Victoria?).

NOTE: Incorrect original spelling: "*Elmis metallicus*".

quatuormaculatus (King, 1865)

Limnius quatuormaculatus King, 1865: 161

TYPE LOCALITY: Australia (New South Wales).

DISTRIBUTION: Australia (ACT, New South Wales).

NOTE: Incorrect original spelling: "*quatuor-maculatus*"; incorrect subsequent spelling: "*quatuor maculatus*" (Carter & Zeck 1929: 55).

tinctus Carter & Zeck, 1929

Kingolus tinctus Carter & Zeck, 1929: 57

TYPE LOCALITY: Australia (New South Wales).

DISTRIBUTION: Australia (New South Wales).

tyrrhenus Carter & Zeck, 1929

Kingolus tyrrhenus Carter & Zeck, 1929: 56

TYPE LOCALITY: Australia (New South Wales).

DISTRIBUTION: Australia (New South Wales).

yarrensis Carter & Zeck, 1929

Kingolus yarrensis Carter & Zeck, 1929: 57

TYPE LOCALITY: Australia (Victoria).

DISTRIBUTION: Australia (Victoria).

Laorina Jäch, 1997

Laorina Jäch, 1997: 393. — Type species: *Laorina schillhammeri* Jäch, 1997. — Gender feminine.

schillhammeri Jäch, 1997

Laorina schillhammeri Jäch, 1997: 396

Type Locality: Laos.

Distribution: China (Yunnan), India (Uttarakhand), Laos, Myanmar.

Lara Le Conte, 1852

Lara Le Conte, 1852: 42. — Type species: *Lara avara* Le Conte, 1852. — Gender feminine.

Note: The name *Lara* has already been used by Drapiez (1819: 54) for a genus of Hymenoptera (Sphecidae), but this name must be regarded as an incorrect subsequent spelling of *Larra* Fabricius, 1793: 220 (Hymenoptera: Sphecidae) and is therefore unavailable (see Spangler 1986, 1987); erroneously, Neave (1939: 868) listed *Lara* Drapiez, 1819 as an available name, and Özdikmen (2008: 39) erroneously regarded it as a synonym of *Larra* Fabricius, 1793.

avara amplipennis Darlington, 1929

Lara avara amplipennis Darlington, 1929: 330

Type Locality: USA (Washington).

Distribution: USA (Washington).

Note: Might be a junior synonym of *Lara avara avara* Le Conte, 1852.

avara avara Le Conte, 1852

Lara avara Le Conte, 1852: 42

Type Locality: USA (California).

Distribution: Canada (British Columbia), USA (California, Colorado, Idaho, Montana, Oregon, Utah, Washington, Wyoming).

gehringi Darlington, 1929

Lara gehringi Darlington, 1929: 329

Type Locality: USA (Washington).

Distribution: USA (California, Oregon, Washington).

Note: Might be a junior synonym of *Lara avara avara* Le Conte, 1852 (Shepard 1993: 2).

Lathridelmis Delève, 1965a

Lathridelmis Delève, 1965a: 105. — Type species: *Lathridelmis crenicollis* Delève, 1965a. — Gender feminine.

crenicollis Delève, 1965a

Lathridelmis crenicollis Delève, 1965a: 106

Type Locality: Congo (DR).

Distribution: Congo (DR).

Leielmis Delève, 1964b

Leielmis Delève, 1964b: 159. — Type species: *Helmis georyssoides* Grouvelle, 1890. — Gender feminine.

georyssoides (Grouvelle, 1890)

Helmis georyssoides Grouvelle, 1890: CCXII[212]

Type Locality: South Africa.

Distribution: Angola, South Africa.

Lemalelmis Spangler, 1981a

Lemalelmis Spangler, 1981a: 380. — Type species: *Lemalelmis minyops* Spangler, 1981a. — Gender feminine.

Note: *Lemalelmis* is obviously a junior synonym of *Hexacylloepus* Hinton, 1940a (see Spangler 1981a: 386: "These new phreatic taxa are closely related ... to ... *Cylloepus haitianus* [*Hexacylloepus haitianus*] ... and apparently are derived from this ... species").

fontana Spangler, 1981a

Lemalelmis fontana Spangler, 1981a: 383

Type Locality: Haiti.

Distribution: Haiti.

minyops Spangler, 1981a

Lemalelmis minyops Spangler, 1981a: 381

Type Locality: Haiti.

Distribution: Haiti.

Leptelmis Sharp, 1888

Leptelmis Sharp, 1888: 243. — Type species: *Leptelmis gracilis* Sharp, 1888. — Gender feminine.

Note: Incorrect subsequent spelling: "*Lephthelmis*" (Zaitzev 1908: 299), not to be regarded as an emendation (ICZN 1999: Art. 33.2.1).

Synonym: *Lephthelmis* Zaitzev, 1910: 21. — Type species: *Leptelmis gracilis* Sharp, 1888. — Gender feminine. Note: Unjustified emendation of *Leptelmis* Sharp, 1888 (ICZN 1999: Art. 33.2).

amoena Delève, 1966c

Leptelmis amoena Delève, 1966c: 8

Type Locality: Ivory Coast.

Distribution: Ghana, Ivory Coast.

basalis Delève, 1968c

Leptelmis basalis Delève, 1968c: 155

Type Locality: Vietnam.

Distribution: Vietnam.

brunnelineata Zhang, Su & Yang, 2003b

Leptelmis brunnelineata Zhang, Su & Yang, 2003b: 189

Type Locality: China (Guangxi).

Distribution: China (Guangxi).

cederholmi Delève, 1973a

Leptelmis cederholmi Delève, 1973a: 11

Type Locality: Sri Lanka.

Distribution: Sri Lanka.

collarti (Delève, 1937a)

Stenelmis collarti Delève, 1937a: 152

Type Locality: Congo (DR).

Distribution: Congo (DR), Congo (R), Gabon.

costulata Delève, 1942

Leptelmis costulata Delève, 1942: 1

Type Locality: Congo (DR).

Distribution: Congo (DR), Congo (R).

Note: Incorrect original spelling: "*costulatus*".

flavicollis (Bollow, 1941)

Lephthelmis flavicollis Bollow, 1941: 86

Type Locality: "South China".

Distribution: "South China".

formosana formosana Nomura, 1962: 48

Leptelmis formosana Nomura, 1962: 48

Type Locality: Taiwan.

DISTRIBUTION: Taiwan.

NOTE: Incorrect subsequent spelling: "*formosanus*" (Hua 2002: 98).

formosana vietnamensis Delève, 1968c

Leptelmis formosana vietnamensis Delève, 1968c: 154

TYPE LOCALITY: Vietnam.

DISTRIBUTION: Vietnam.

fracticollis Champion, 1923

Leptelmis fracticollis Champion, 1923: 167

TYPE LOCALITY: India (Uttarakhand).

DISTRIBUTION: India (Uttarakhand).

fragilis Delève, 1966c

Leptelmis fragilis Delève, 1966c: 7

TYPE LOCALITY: South Africa.

DISTRIBUTION: South Africa.

fujiana Yang & Zhang, 2002

Leptelmis fujiana Yang & Zhang, 2002: 812

TYPE LOCALITY: China (Fujian).

DISTRIBUTION: China (Fujian).

gracilis gracilis Sharp, 1888

Leptelmis gracilis Sharp, 1888: 244

TYPE LOCALITY: Japan (Honshu).

DISTRIBUTION: Japan (Honshu, Kyushu, Shikoku), Korea (R).

SYNONYMS: *Leptelmis parallela* Nomura, 1962: 46. — Type locality: Japan (Shikoku). NOTE: Synonymized by Hayashi et al. (2013: 693).

Leptelmis coreana Jung & Bae, 2012: 255. — Type locality: Korea (R).

Leptelmis ochra Jung & Bae, 2012: 256. — Type locality: Korea (R).

gracilis impubis Zhang & Ding, 1995

Leptelmis gracilis impubis Zhang & Ding, 1995: 17

TYPE LOCALITY: China (Beijing).

DISTRIBUTION: China (Beijing).

guangxiana Zhang, Su & Yang, 2003b

Leptelmis guangxiana Zhang, Su & Yang, 2003b: 191

Type Locality: China (Guangxi).

Distribution: China (Guangxi).

Note: Similar to *Leptelmis gracilis* Sharp, 1888 (Jung et al. 2015: 103).

major Delève, 1966c

Leptelmis major Delève, 1966c: 4

Type Locality: Congo (DR).

Distribution: Congo (DR).

obscura Delève, 1968c

Leptelmis obscura Delève, 1968c: 153

Type Locality: Vietnam.

Distribution: China (Guangxi) (Zhang, Su & Yang 2003b: 192), Vietnam.

orchymonti Delève, 1942

Leptelmis orchymonti Delève, 1942: 3

Type Locality: Congo (DR).

Distribution: Congo (DR).

philomina Brown & Thobias, 1984

Leptelmis philomina Brown & Thobias, 1984: 24

Type Locality: India ("Western Ghats").

Distribution: India ("Western Ghats").

seydeli Delève, 1966c

Leptelmis seydeli Delève, 1966c: 6

Type Locality: Congo (DR).

Distribution: Congo (DR).

signata Delève, 1968c

Leptelmis signata Delève, 1968c: 152

Type Locality: Vietnam.

Distribution: Vietnam.

sobrina Delève, 1974

Leptelmis sobrina Delève, 1974: 272

Type Locality: Ghana.

Distribution: Ghana.

stricticollis (Grouvelle, 1896a)

Stenelmis stricticollis Grouvelle, 1896a: 45

TYPE LOCALITY: Indonesia (Sumatra).
DISTRIBUTION: Indonesia (Sumatra).

sulcata (Grouvelle, 1892a)

Stenelmis sulcata Grouvelle, 1892a: 188
TYPE LOCALITY: Indonesia (Sumatra).
DISTRIBUTION: Indonesia (Sumatra).

tawitawiensis Delève, 1973c

Leptelmis tawitawiensis Delève, 1973c: 28
TYPE LOCALITY: Philippines (Tawi-Tawi Islands).
DISTRIBUTION: Philippines (Tawi-Tawi Islands).

vittata Zhang, Su & Yang, 2003b

Leptelmis vittata Zhang, Su & Yang, 2003b: 190
TYPE LOCALITY: China (Guangxi).
DISTRIBUTION: China (Guangxi).

Limnius Illiger, 1802

Limnius Illiger, 1802: 297. — Type species: *Dytiscus volckmari* Panzer, 1793. — Gender masculine.
NOTE: Incorrect subsequent spelling: "*Lymnius*" (Bertolini 1874: 102).
SYNONYMS: *Latelmis* Reitter, 1883: 75. — Type species: *Dytiscus volckmari* Panzer, 1793. — Gender feminine.
Lathelmis Zaitzev, 1908: 307. — Type species: *Dytiscus volckmari* Panzer, 1793. — Gender feminine. NOTE: Unjustified emendation of *Latelmis* Reitter, 1883 (ICZN 1999: Art. 33.2.1).

colchicus Delève, 1963b

Limnius colchicus Delève, 1963b: 7
TYPE LOCALITY: Turkey.
DISTRIBUTION: Armenia, Bulgaria, Georgia, Greece, Turkey.

gibbosus Jäch, 1984b

Limnius gibbosus Jäch, 1984b: 35
TYPE LOCALITY: Turkey.
DISTRIBUTION: Turkey.

intermedius Fairmaire, 1881

Limnius intermedius Fairmaire, 1881: XI[11]
TYPE LOCALITY: Italy (Sardinia).

Distribution: Albania, Algeria, Armenia, Austria, Belarus, Bosnia and Herzegovina, Bulgaria, Croatia, Czech Republic, Denmark, France, (Germany), Greece, Hungary, Israel, Italy, Lebanon, Macedonia, (Montenegro), Morocco, Palestine (West Bank), Poland? (Przewoźny et al. 2011), Portugal, Romania, Russia, Serbia, Slovakia, Slovenia, Spain, Tunisia, Turkey, Ukraine; the record of *L. muelleri* from Lithuania by Tamutis et al (2011: 215) and other authors cited therein might well refer to this species.

Synonyms: *Latelmis opaca subcrenata* Rey, 1889: 67. — Type locality: France. Note: Originally described as a variety of *Latelmis opaca*; synonymized with *L. intermedius* by Berthélemy (1964b: 527).

Elmis rufiventris Kuwert, 1890: 45. — Type locality: Croatia or Bosnia and Herzegovina (no lectotype designated). Note: Synonymized with *L. intermedius* by Berthélemy (1979: 47).

Elmis palaestina Pic, 1901: 3. — Type locality: Palestine (West Bank). Note: Incorrect original spelling: "*Elmis palæstinus*"; synonymized with *L. intermedius* by Berthélemy (1979: 47).

Elmis longior Pic, 1905: 145. — Type locality: Algeria. Note: Formally, this name is a junior secondary homonym of *Elmis longior* (Grouvelle, 1896a) (see below, under "Species incertae sedis"), however, this name has not been replaced before 1961 and the relevant taxa are not considered to be congeneric, therefore, the name *Elmis longior* Pic, 1905 remains valid (ICZN 1999: Art. 59.3); synonymized with *L. intermedius* by Berthélemy (1964a: 265).

Lathelmis horioni Illies, 1953: 179. — Type locality: Czech Republic. Note: Synonymized with *L. intermedius* (and regarded as a possible subspecies) by Olmi (1976: 188–189).

letourneuxi (Pic, 1894)

Elmis letourneuxi Pic, 1894: 195

Type Locality: Syria.

Distribution: Israel, Syria.

muelleri (Erichson, 1847)

Elmis muelleri Erichson, 1847: 529

Type Locality: Austria, France, or Germany (obviously no lectotype designated); syntypes recorded from Italy (Sardinia) should belong to *Limnius intermedius*.

Distribution: Austria, Czech Republic, France, Germany, Hungary, Switzerland; probably extant in France only; the record from Croatia (Imotski) by Novak (1952: 151) is probably based on a misidentification; the record from Lithuania by Tamutis et al (2011: 215) and other authors cited therein is rather doubtful.

Note: Incorrect original spelling: "*mülleri*"; named after German coleopterist Philipp Wilbrand Jacob Müller (1771–1851); incorrect subsequent spelling: "*mulleri*" (Bertrand 1972: 491).

Synonym: *Latelmis oblonga* Rey, 1889: 67. — Type locality: France. Note: Synonymized with *L. muelleri* by Berthélemy (1964b: 526).

opacus Müller, 1806a

Limnius opacus Müller, 1806a: 197

Type Locality: Germany.

Distribution: Algeria, Armenia, Austria, Belgium, Bosnia and Herzegovina, Bulgaria, Croatia, Czech Republic, France, Germany, Greece, Hungary, Israel, Italy, Lebanon, Luxembourg, (Montenegro), Morocco, Netherlands, Poland? (Przewoźny et al. 2011), Portugal, Romania, Russia, Serbia, Slovakia, Slovenia, Spain, Switzerland, Turkey, Ukraine; the record from Latvia by Telnov (2004: 62) is based on misidentification (D. Telnov, pers. comm.).

Synonyms: *Elmis lepidoptera* Kuwert, 1890: 45. — Type locality: Bosnia and Herzegovina. Note: Incorrect original spelling: "*Elmis lepidopterus*".

Elmis carreti Pic, 1895: CXXVIII[128]. — Type locality: Algeria.

Latelmis subopaca Ganglbauer, 1904a: 118. — Type locality: Italy.

Latelmis carreti dissidens Alluaud, 1922: 41. — Type locality: Morocco.

Latelmis liouvillei Alluaud, 1922: 42. — Type locality: Morocco. Note: Incorrect subsequent spelling: "*lionvillei*" (Winkler 1926: column 672).

Latelmis jahandiezi Alluaud, 1922: 42. — Type locality: Morocco. Note: Regarded as a possible subspecies of *Limnius opacus* by Berthélemy (1964a: 265).

Lathelmis latiuscula Zaitzev, 1947: 90 **syn.n.** — Type locality: Armenia. Note: Incorrect original spelling: "*latiusculus*".

perrisi carinatus (Perez Arcas, 1865)

Elmis carinata Perez Arcas, 1865: 413

Type Locality: Spain.

Distribution: Portugal, Spain.

Note: Incorrect original spelling: "*Elmis carinatus*".

Synonyms: *Elmis subcarinatus* Sharp, 1872: 263. — Type locality: Spain.

Limnius perrisi mediocarinatus Berthélemy, 1964a: 261. — Type locality: Spain.

perrisi perrisi (Dufour, 1843)

Elmis perrisi Dufour, 1843: 56

Type Locality: France.

Distribution: Andorra, Albania, Austria, Belgium, Bosnia and Herzegovina, Bulgaria, Croatia, Czech Republic, France, Germany, Greece, Hungary, Italy, Luxembourg, Macedonia, (Montenegro), Netherlands, Poland, Romania, Serbia, Slovakia, Slovenia, Spain, Switzerland, Turkey, Ukraine.

Note: The page number of the original description erroneously was cited as 317 (which is actually the species number) by Zaitzev (1908: 308, 1910: 35).

Synonyms: *Elmis germari* Erichson, 1847: 528. — Type locality: Germany.

Latelmis rambouseki Mařan, 1939: 41 **syn.n.** — Type locality: Macedonia.

satanus Jäch, 1982b

Limnius satanus Jäch, 1982b: 93

Type Locality: Turkey.

Distribution: Turkey.

stygius Hernando, Aguilera & Ribera, 2001

Limnius stygius Hernando, Aguilera & Ribera, 2001: 69

TYPE LOCALITY: Morocco.

DISTRIBUTION: Morocco.

sulcipennis damryi Fairmaire, 1881

Limnius damryi Fairmaire, 1881: XII[12]

TYPE LOCALITY: France (Corsica).

DISTRIBUTION: France (Corsica).

sulcipennis sulcipennis Fairmaire, 1881

Limnius sulcipennis Fairmaire, 1881: XII[12]

TYPE LOCALITY: Italy (Sardinia).

DISTRIBUTION: Italy (Sardinia).

surcoufi (Pic, 1905)

Elmis surcoufi Pic, 1905: 145

TYPE LOCALITY: Algeria.

DISTRIBUTION: Algeria.

swayambhu Jäch, 1982b

Limnius swayambhu Jäch, 1982b: 92

TYPE LOCALITY: Nepal.

DISTRIBUTION: Nepal.

volckmari (Panzer, 1793)

Dytiscus volckmari Panzer, 1793: 4

TYPE LOCALITY: Germany.

DISTRIBUTION: Armenia, Austria, Belarus, Belgium, Bosnia and Herzegovina, Bulgaria, Croatia, Czech Republic, Denmark, Estonia, Finland, France, Germany, Greece, Hungary, Iran, Ireland, Italy, Latvia, Lithuania, Luxembourg, Macedonia, Montenegro (V. Mičetić Stanković, pers. comm.), Netherlands, Norway, Poland, Portugal, Romania, Russia, Serbia, Slovakia, Slovenia, Spain, Sweden, Switzerland, Turkey, Ukraine, United Kingdom.

NOTE: Incorrect subsequent spellings: "*volkmari*" (Illiger 1802: 297, Marsham 1802: 192, Erichson 1847: 527), "*volckmarii*" (Villa & Villa 1833: 14), "*volkmarii*" (Motschulsky 1853: 13).

SYNONYMS: *Chrysomela buprestoides* Marsham, 1802: 192 (Chrysomelidae). — Type locality: United Kingdom. NOTE: Synonymy by Samouelle (1819: 185), but listed as a doubtful synonym (with question mark) by Zaitzev (1910: 36); types not traced in Natural History Museum, London, U.K. (R. Booth, email of 23.VI.2014).

Latelmis cecconii Reitter, 1906: 239. — Type locality: Italy.

wewalkai Jäch, 1984b

Limnius wewalkai Jäch, 1984b: 36

TYPE LOCALITY: Turkey.

DISTRIBUTION: Turkey.

Lobelmis Fairmaire, 1898b

Lobelmis Fairmaire, 1898b: 467. — Type species: *Lobelmis cucullata* Fairmaire, 1898b. — Gender feminine.

NOTE: Incorrect subsequent spelling: "*Lophelmis*" Zaitzev (1908: 310), not to be regarded as an emendation (ICZN 1999: Art. 33.2.1).

SYNONYM: *Lophelmis* Zaitzev, 1910: 37. — Type species: *Lobelmis cucullata* Fairmaire, 1898b. — Gender feminine. NOTE: Unjustified emendation of *Lobelmis* Fairmaire, 1898b (ICZN 1999: Art. 33.2.1).

cucullata cucullata Fairmaire, 1898b

Lobelmis cucullata Fairmaire, 1898b: 467

TYPE LOCALITY: Madagascar.

DISTRIBUTION: Madagascar.

NOTE: Incorrect original spelling: "*cucullatus*".

cucullata tanalana Delève, 1964a

Lobelmis cucullata tanalana Delève, 1964a: 40

TYPE LOCALITY: Madagascar.

DISTRIBUTION: Madagascar.

harrisoni Delève, 1967c

Lobelmis harrisoni Delève, 1967c: 81

TYPE LOCALITY: Zimbabwe.

DISTRIBUTION: Zimbabwe.

lineicollis (Fairmaire, 1902)

Limnius lineicollis Fairmaire, 1902: 344

TYPE LOCALITY: Madagascar.

DISTRIBUTION: Madagascar.

minuta Delève, 1964a

Lobelmis minuta Delève, 1964a: 42

TYPE LOCALITY: Madagascar.

DISTRIBUTION: Madagascar.

odiosa (Grouvelle, 1906a)

Elmidolia odiosa Grouvelle, 1906a: 156

Type Locality: Madagascar.

Distribution: Madagascar.

subnigra Grouvelle, 1906c

Lobelmis subnigra Grouvelle, 1906c: 318

Type Locality: Tanzania.

Distribution: Tanzania.

vicina Delève, 1964a

Lobelmis vicina Delève, 1964a: 41

Type Locality: Madagascar.

Distribution: Madagascar.

Loxostirus Jäch & Kodada, 1996b

Loxostirus Jäch & Kodada, 1996b: 406. — Type species: *Loxostirus willi* Jäch & Kodada, 1996b. — Gender masculine.

willi Jäch & Kodada, 1996b

Loxostirus willi Jäch & Kodada, 1996b: 408

Type Locality: Malaysia (Sarawak).

Distribution: Malaysia (Sarawak).

zelenka Jäch & Kodada, 1996b

Loxostirus zelenka Jäch & Kodada, 1996b: 410

Type Locality: Malaysia (Sarawak).

Distribution: Malaysia (Sarawak).

Luchoelmis Spangler & Staines, 2004a

Luchoelmis Spangler & Staines, 2004a: 215. — Type species: *Luchoelmis penai* Spangler & Staines, 2004a. — Gender feminine.

aequalis Spangler & Staines, 2004a

Luchoelmis aequalis Spangler & Staines, 2004a: 216

Type Locality: Chile.

Distribution: Chile.

cekalovici Spangler & Staines, 2004a

Luchoelmis cekalovici Spangler & Staines, 2004a: 217

TYPE LOCALITY: Chile.

DISTRIBUTION: Argentina, Chile.

kapenkemkensis Archangelsky & Brand, 2014

Luchoelmis kapenkemkensis Archangelsky & Brand, 2014: 564

TYPE LOCALITY: Argentina.

DISTRIBUTION: Argentina.

magallanensis Spangler & Staines, 2004a

Luchoelmis magallanensis Spangler & Staines, 2004a: 218

TYPE LOCALITY: Chile.

DISTRIBUTION: Chile.

penai Spangler & Staines, 2004a

Luchoelmis penai Spangler & Staines, 2004a: 219

TYPE LOCALITY: Chile.

DISTRIBUTION: Chile.

Ludyella Reitter, 1899

Ludyella Reitter, 1899: 283. — Type species: *Ludyella corticariiformis* Reitter, 1899. — Gender feminine.

NOTE: *Ludyella* might be a senior synonym of one of the Afrotropical genera, see Jäch (1983b).

corticariiformis Reitter, 1899

Ludyella corticariiformis Reitter, 1899: 284

TYPE LOCALITY: Unknown, probably in the Afrotropical Region (see Jäch 1983b).

DISTRIBUTION: Unknown, probably Afrotropical (see Jäch 1983b).

Macrelmis Motschulsky, 1860

Macrelmis Motschulsky, 1860: 52. — Type species: *Macrelmis dentata* Motschulsky, 1860. — Gender feminine.

NOTE: Incorrect subsequent spelling: "*Machrhelmis*" Zaitzev (1908: 311), not to be regarded as an emendation (ICZN 1999: Art. 33.2.1).

SYNONYMS: *Elsianus* Sharp, 1882: 131. — Type species: *Elsianus striatus* Sharp, 1882, designated by Hinton (1940a: 263); Brown (1983: 5) erroneously regarded *Elsianus robustus* Sharp, 1882 as the type species of *Elsianus*. — Gender masculine.

Uralohelmis Roubal, 1940: 151. — Type species: *Uralohelmis jureceki* Roubal, 1940. — Gender feminine. Note: Synonymized by Jäch (2002: 163).

aeolus (Hinton, 1946b)

Elsianus aeolus Hinton, 1946b: 139

Type Locality: Brazil (Santa Catarina).

Distribution: Brazil (Santa Catarina), Peru.

Note: Epithet name is a noun in apposition (ruler of the winds in Greek mythology); incorrect subsequent spelling: "*aeolis*" (Brown 1984: 124).

aequalis (Hinton, 1937e)

Elsianus aequalis Hinton, 1937e: 97

Type Locality: Brazil (Santa Catarina).

Distribution: Argentina, Brazil (Santa Catarina).

aleus (Hinton, 1946b)

Elsianus aleus Hinton, 1946b: 138

Type Locality: Brazil (Santa Catarina).

Distribution: Brazil (Santa Catarina).

Note: Epithet name is a noun in apposition (king of Arcadia); incorrect subsequent spelling: "*alea*" (Brown 1984: 124).

amanus (Hinton, 1946b)

Elsianus amanus Hinton, 1946b: 137

Type Locality: Brazil (Santa Catarina).

Distribution: Brazil (Santa Catarina), Paraguay.

Note: Epithet name probably is a noun in apposition; incorrect subsequent spelling: "*amana*" (Brown 1984: 124).

amazonica (Hinton, 1945b)

Elsianus amazonicus Hinton, 1945b: 91

Type Locality: Brazil (Rondônia); erroneously, Hinton (1945b: 92) attributed the type locality to Mato Grosso ("Matto [sic] Grosso: Porto Velho, i.ix.1937 (H.E. Hinton)").

Distribution: Brazil (Rondônia), Panama, Peru; erroneously recorded from Mato Grosso by Segura et al. (2013: 25).

antiqua (Grouvelle, 1896a)

Elsianus antiquus Grouvelle, 1896a: 41

Type Locality: Brazil (details unknown).

Distribution: Brazil (details unknown).

aristaeus (Hinton, 1946b)

Elsianus aristaeus Hinton, 1946b: 134

TYPE LOCALITY: Brazil (Santa Catarina).

DISTRIBUTION: Brazil (Santa Catarina).

NOTE: Epithet name is a noun in apposition (son of Apollo and Cyrene); incorrect subsequent spelling: "*aristea*" (Brown 1984: 124).

bicolor (Hinton, 1939e)

Elsianus bicolor Hinton, 1939e: 35

TYPE LOCALITY: Bolivia.

DISTRIBUTION: Bolivia.

bispo Barbosa, Fernandes & Oliveira, 2013

Macrelmis bispo Barbosa, Fernandes & Oliveira, 2013: 130

TYPE LOCALITY: Brazil (Goiás).

DISTRIBUTION: Brazil (Goiás).

celsus (Hinton, 1946b)

Elsianus celsus Hinton, 1946b: 145

TYPE LOCALITY: Brazil (Santa Catarina).

DISTRIBUTION: Brazil (Santa Catarina).

NOTE: Epithet name is a noun in apposition (Greek philosopher); incorrect subsequent spelling: "*celsa*" (Brown 1984: 124).

clypeata (Hinton, 1936c)

Elsianus clypeatus Hinton, 1936c: 424

TYPE LOCALITY: Trinidad and Tobago (Trinidad).

DISTRIBUTION: Trinidad and Tobago (Tobago, Trinidad), Venezuela.

codrus (Hinton, 1946b)

Elsianus codrus Hinton, 1946b: 147

TYPE LOCALITY: Brazil (Santa Catarina).

DISTRIBUTION: Brazil (Santa Catarina).

NOTE: Epithet name is a noun in apposition (king of Athens); incorrect subsequent spelling: "*codris*" (Brown 1984: 124).

coquereli (Grouvelle, 1889b)

Elsianus coquereli Grouvelle, 1889b: 396

TYPE LOCALITY: Colombia.

DISTRIBUTION: Colombia.

dentata Motschulsky, 1860

Macrelmis dentata Motschulsky, 1860: 52

TYPE LOCALITY: Colombia.

DISTRIBUTION: Colombia.

NOTE: Incorrect original spelling: "*dentatus*".

elicioi Monte & Mascagni, 2012

Macrelmis elicioi Monte & Mascagni, 2012: 21

TYPE LOCALITY: Ecuador.

DISTRIBUTION: Ecuador.

fluminensis Sampaio, Passos & Ferreira, 2012

Macrelmis fluminensis Sampaio, Passos & Ferreira, 2012: 164

TYPE LOCALITY: Brazil (Rio de Janeiro).

DISTRIBUTION: Brazil (Rio de Janeiro).

friburguensis Sampaio, Passos & Ferreira, 2012

Macrelmis friburguensis Sampaio, Passos & Ferreira, 2012: 166

TYPE LOCALITY: Brazil (Rio de Janeiro).

DISTRIBUTION: Brazil (Rio de Janeiro).

NOTE: Incorrect subsequent spelling: "*friburgensis*" (Segura et al. 2013: 26).

froehlichi Barbosa, Fernandes & Oliveira, 2013

Macrelmis froehlichi Barbosa, Fernandes & Oliveira, 2013: 132

TYPE LOCALITY: Brazil (Goiás).

DISTRIBUTION: Brazil (Goiás).

germaini (Grouvelle, 1896a)

Cylloepus germaini Grouvelle, 1896a: 43

TYPE LOCALITY: Bolivia.

DISTRIBUTION: Bolivia.

grandis (Hinton, 1934)

Elsianus grandis Hinton, 1934: 195

TYPE LOCALITY: Mexico.

DISTRIBUTION: Mexico.

graniger (Sharp, 1882)

Elsianus graniger Sharp, 1882: 133

TYPE LOCALITY: Costa Rica.

DISTRIBUTION: Costa Rica, Guatemala, Mexico, Nicaragua, Peru.

NOTE: Incorrect subsequent spelling: "*granigera*" (Brown 1984: 125) — see ICZN (1999: Art. 31.2.2).

grouvellei (Hinton, 1936f)

Cylloepus grouvellei Hinton, 1936f: 55

TYPE LOCALITY: Brazil (Rio de Janeiro).

DISTRIBUTION: Brazil (Rio de Janeiro).

NOTE: Substitute name for *Cylloepus granosus* Grouvelle, 1896a.

HOMONYM: *Cylloepus granosus* Grouvelle, 1896a: 44. NOTE: Formerly a junior secondary homonym of *Helmis granosa* Grouvelle, 1889b (*Hexacylloepus*); permanently invalid (ICZN 1999: Art. 59.3).

hayekae Spangler, 1997

Macrelmis hayekae Spangler, 1997: 2

TYPE LOCALITY: Venezuela.

DISTRIBUTION: Venezuela.

immsi (Hinton, 1936c)

Elsianus immsi Hinton, 1936c: 418

TYPE LOCALITY: Costa Rica.

DISTRIBUTION: Costa Rica.

isus (Hinton, 1946b)

Elsianus isus Hinton, 1946b: 129

TYPE LOCALITY: Brazil (Santa Catarina).

DISTRIBUTION: Argentina, Bolivia, Brazil (Goiás, Rio de Janeiro, Santa Catarina), Paraguay.

NOTE: Epithet name is a noun in apposition (son of Priam in Greek mythology); incorrect subsequent spelling: "*isis*" (Brown 1984: 125, Sampaio et al. 2012: 164).

jureceki (Roubal, 1940)

Uralohelmis jureceki Roubal, 1940: 152

TYPE LOCALITY: "Russia (Orenburg)" according to original description, but probably in USA (see Jäch 2002).

DISTRIBUTION: Russia?, USA?

NOTE: Original description in Slovak (p. 152) and Latin (p. 154).

leonilae Spangler & Santiago F.[Fragoso], 1986

Macrelmis leonilae Spangler & Santiago F.[Fragoso], 1986: 155

TYPE LOCALITY: Mexico.

DISTRIBUTION: Costa Rica, Guatemala, Mexico.

milleri Spangler, 1997

Macrelmis milleri Spangler, 1997: 1

Type Locality: Venezuela.

Distribution: Venezuela.

moesta (Horn, 1870)

Elmis moesta Horn, 1870: 37

Type Locality: USA (Arizona).

Distribution: USA (Arizona).

Note: Incorrect original spelling: "*Elmis moestus*".

nessimiani Barbosa, Fernandes & Oliveira, 2013

Macrelmis nessimiani Barbosa, Fernandes & Oliveira, 2013: 135

Type Locality: Brazil (Goiás).

Distribution: Brazil (Goiás).

peruviana (Hinton, 1945b)

Elsianus peruvianus Hinton, 1945b: 90

Type Locality: Peru.

Distribution: Peru.

plaumanni (Hinton, 1946b)

Elsianus plaumanni Hinton, 1946b: 143

Type Locality: Brazil (Santa Catarina).

Distribution: Brazil (Santa Catarina), Paraguay.

pubescens (Grouvelle, 1889b)

Elsianus pubescens Grouvelle, 1889b: 396

Type Locality: Brazil (Santa Catarina).

Distribution: Brazil (Santa Catarina).

robusta (Sharp, 1882)

Elsianus robustus Sharp, 1882: 132

Type Locality: Guatemala.

Distribution: Guatemala.

saleius (Hinton, 1946b)

Elsianus saleius Hinton, 1946b: 144

Type Locality: Brazil (Santa Catarina).

Distribution: Brazil (Santa Catarina).

Note: Epithet name is a noun in apposition (Roman poet); incorrect subsequent spelling: "*saleia*" (Brown 1984: 125).

saltensis Manzo, 2003

Macrelmis saltensis Manzo, 2003: 170

Type Locality: Argentina.

Distribution: Argentina, Bolivia.

salti (Hinton, 1936b)

Elsianus salti Hinton, 1936b: 283

Type Locality: Brazil (São Paulo).

Distribution: Brazil (São Paulo).

sandersoni (Hinton, 1936c)

Elsianus sandersoni Hinton, 1936c: 420

Type Locality: Mexico.

Distribution: Mexico.

scutellaris (Hinton, 1934)

Elsianus scutellaris Hinton, 1934: 196

Type Locality: Mexico.

Distribution: Costa Rica, Mexico.

shoemakei (Brown, 1971)

Elsianus shoemakei Brown, 1971: 56

Type Locality: USA (Texas).

Distribution: Mexico, USA (Texas).

striata (Sharp, 1882)

Elsianus striatus Sharp, 1882: 132

Type Locality: Guatemala.

Distribution: Costa Rica, Guatemala, Mexico, Panama, Peru.

striatoides (Hinton, 1936c)

Elsianus striatoides Hinton, 1936c: 422

Type Locality: Mexico.

Distribution: Mexico.

Note: Incorrect subsequent spelling: "*striatoidea*" (Brown 1984: 125).

tarsalis (Hinton, 1936c)

Elsianus tarsalis Hinton, 1936c: 419

TYPE LOCALITY: Costa Rica.

DISTRIBUTION: Costa Rica, Ecuador, Panama.

tereus (Hinton, 1946b)

Elsianus tereus Hinton, 1946b: 134

TYPE LOCALITY: French Guiana.

DISTRIBUTION: French Guiana.

NOTE: Epithet name is a noun in apposition (a king of Thrace); incorrect subsequent spelling: "*terea*" (Brown 1984: 125).

texana (Schaeffer, 1911)

Elsianus texanus Schaeffer, 1911: 119

TYPE LOCALITY: USA (Texas).

DISTRIBUTION: Mexico, USA (Texas).

NOTE: Incorrect subsequent spelling: "*Macrelmis texanus*" (Poole & Gentili 1996: 264).

thorpei (Hinton, 1946b)

Elsianus thorpei Hinton, 1946b: 131

TYPE LOCALITY: Brazil (Amazonas).

DISTRIBUTION: Brazil (Amazonas).

tijucana Passos & Felix, 2004a

Macrelmis tijucana Passos & Felix, 2004a: 49

TYPE LOCALITY: Brazil (Rio de Janeiro).

DISTRIBUTION: Brazil (Rio de Janeiro).

tityrus (Hinton, 1946b)

Elsianus tityrus Hinton, 1946b: 133

TYPE LOCALITY: Brazil (Amazonas).

DISTRIBUTION: Brazil (Amazonas).

NOTE: Epithet name is a noun in apposition and may refer to various figures in the Greek mythology (e.g. an ovine beast, or the father of the Amazon Helene); incorrect subsequent spelling: "*tityra*" (Brown 1984: 126).

zamus (Hinton, 1946b)

Elsianus zamus Hinton, 1946b: 141

TYPE LOCALITY: Brazil (Santa Catarina).

DISTRIBUTION: Brazil (Santa Catarina).

NOTE: Epithet name might be a noun in apposition (ICZN 1999: Art. 31.2.2); incorrect subsequent spelling: "*zama*" (Brown 1984: 126).

Macronevia Jäch & Boukal, 1996

Macronevia Jäch & Boukal, 1996: 179. — Type species: *Macronychus simplex* Hinton, 1936c. — Gender feminine.

simplex (Hinton, 1936c)

Macronychus simplex Hinton, 1936c: 433

TYPE LOCALITY: Malaysia (Selangor).

DISTRIBUTION: Malaysia (Kedah, Kelantan, Pahang, Perak, Selangor), Thailand.

Macronychoides Champion, 1923

Macronychoides Champion, 1923: 171. — Type species: *Macronychoides amnicola* Champion, 1923. — Gender masculine.

amnicola Champion, 1923

Macronychoides amnicola Champion, 1923: 172

TYPE LOCALITY: India (Tamil Nadu).

DISTRIBUTION: India (Tamil Nadu).

Macronychus Müller, 1806b

Macronychus Müller, 1806b: 207. — Type species: *Macronychus quadrituberculatus* Müller, 1806b. — Gender masculine.

NOTE: Incorrect subsequent spelling: "*Macronycus*" (Rafinesque 1815: 112).

glabratus Say, 1825

Macronychus glabratus Say, 1825: 187

TYPE LOCALITY: USA (details unknown, type material not located).

DISTRIBUTION: Canada (New Brunswick, Nova Scotia, Ontario, Quebec), USA (Alabama, Arkansas, Connecticut, Delaware, District of Columbia, Florida, Georgia, Illinois, Indiana, Iowa, Kansas, Kentucky, Louisiana, Maine, Maryland, Minnesota, Mississippi, Missouri, New Jersey, New York, North Carolina, North Dakota, Ohio, Oklahoma, Pennsylvania, Rhode Island, South Carolina, Tennessee, Texas, Virginia, West Virginia, Wisconsin).

NOTE: Incorrect subsequent spelling: "*glabatus*" (Smith 1989: 450, 451).

SYNONYM: *Macronychus lateralis* Melsheimer, 1844: 99. — Type locality: USA (Pennsylvania).

indicus Hinton, 1940f

Macronychus indicus Hinton, 1940f: 118

Type Locality: India (Uttarakhand).

Distribution: India (Uttarakhand), Nepal, Thailand.

jaechi Čiampor & Kodada, 1998

Macronychus jaechi Čiampor & Kodada, 1998: 240

Type Locality: China (Hainan).

Distribution: China (Hainan).

jendeki Čiampor & Kodada, 1998

Macronychus jendeki Čiampor & Kodada, 1998: 245

Type Locality: China (Anhui).

Distribution: China (Anhui, Guangdong, Hunan, Jiangxi).

kubani Čiampor & Kodada, 1998

Macronychus kubani Čiampor & Kodada, 1998: 243

Type Locality: China (Yunnan).

Distribution: China (Sichuan, Yunnan).

levanidovae Lafer, 1980

Macronychus levanidovae Lafer, 1980: 50

Type Locality: Russia (Far East).

Distribution: (Korea (DPR)), Korea (R), Russia (Far East).

quadrituberculatus Müller, 1806b

Macronychus quadrituberculatus Müller, 1806b: 215

Type Locality: Germany.

Distribution: Austria, Belarus, Croatia, Czech Republic, Estonia, Finland, France, Germany, Greece, Hungary, Italy, Latvia, Lithuania, Morocco, Poland, Portugal, Romania, Russia, Serbia, Slovakia, Spain, Switzerland, Ukraine, United Kingdom.

Note: Incorrect original spelling: "*4-tuberculatus*".

Synonym: *Hydrochus parumoculatus* Hardy, 1854: 270 (Hydrochidae). — Type locality: United Kingdom. Note: Synonymized by Bold (1871: 35), see also Zaitzev (1908: 417); synonymized with *Ancyronyx variegatus* (Germar, 1824) by Zaitzev (1910: 40) and Leng (1920: 186) (see also Sanderson 1953a: 159); type not traced in Natural History Museum, London, U.K. (M. Barclay, email of 5.VI.2014).

reticulatus Čiampor & Kodada, 1998

Macronychus reticulatus Čiampor & Kodada, 1998: 237

TYPE LOCALITY: Laos.

DISTRIBUTION: China (Guangdong, Hunan, Yunnan), Laos.

sulcatus Čiampor & Kodada, 1998

Macronychus sulcatus Čiampor & Kodada, 1998: 238

TYPE LOCALITY: Vietnam.

DISTRIBUTION: Vietnam.

ultimus Čiampor & Kodada, 1998

Macronychus ultimus Čiampor & Kodada, 1998: 247

TYPE LOCALITY: Vietnam.

DISTRIBUTION: Laos, Thailand, Vietnam.

vietnamensis Delève, 1968c

Macronychus vietnamensis Delève, 1968c: 179

TYPE LOCALITY: Vietnam.

DISTRIBUTION: Vietnam.

Microcylloepus Hinton, 1935

Microcylloepus Hinton, 1935: 178. — Type species: *Stenelmis pusilla* Le Conte, 1852. — Gender masculine.

acuminatus Bug, 1973

Microcylloepus acuminatus Bug, 1973: 115

TYPE LOCALITY: Brazil (São Paulo).

DISTRIBUTION: Brazil (São Paulo).

angustus Hinton, 1940a

Microcylloepus angustus Hinton, 1940a: 307

TYPE LOCALITY: Mexico.

DISTRIBUTION: Costa Rica, Mexico, Peru.

browni (Hatch, 1938)

Heterelmis browni Hatch, 1938: 16

TYPE LOCALITY: USA (Montana).

DISTRIBUTION: USA (Montana).

carinatus Hinton, 1940a

Microcylloepus carinatus Hinton, 1940a: 304

Type Locality: Trinidad and Tobago (Trinidad).

Distribution: Trinidad and Tobago (Tobago, Trinidad).

Note: Formal description published by Hinton (1945c: 57).

chilensis Janssens, 1957

Microcylloepus chilensis Janssens, 1957: 1

Type Locality: Chile.

Distribution: Chile.

distortus (Sharp, 1882)

Elmis distorta Sharp, 1882: 137

Type Locality: Guatemala.

Distribution: Costa Rica, Guatemala.

Note: Incorrect original spelling: "*Elmis distortus*".

dolon Hinton, 1945c

Microcylloepus dolon Hinton, 1945c: 60

Type Locality: Brazil (Amazonas).

Distribution: Brazil (Amazonas).

femoralis Hinton, 1940h

Microcylloepus femoralis Hinton, 1940h: 241

Type Locality: Brazil (Santa Catarina).

Distribution: Brazil (Santa Catarina), Uruguay.

formicoideus Shepard, 1990

Microcylloepus formicoideus Shepard, 1990: 147

Type Locality: USA (California).

Distribution: USA (California).

grandis Hinton, 1940d

Microcylloepus grandis Hinton, 1940d: 66

Type Locality: Brazil (Santa Catarina).

Distribution: Brazil (Santa Catarina), Paraguay.

granosus Hinton, 1940d

Microcylloepus granosus Hinton, 1940d: 64

Type Locality: Brazil (Santa Catarina).

Distribution: Brazil (Santa Catarina).

immsi (Hinton, 1937e)

Cylloepus immsi Hinton, 1937e: 98

TYPE LOCALITY: Brazil (Santa Catarina).

DISTRIBUTION: Brazil (Santa Catarina).

inaequalis (Sharp, 1882)

Elmis inaequalis Sharp, 1882: 137

TYPE LOCALITY: Guatemala.

DISTRIBUTION: Brazil (Goiás, Santa Catarina), Costa Rica, Guatemala, Mexico, Nicaragua, Panama, Paraguay.

NOTE: Incorrect subsequent spelling: "*inœqualis*" (Grouvelle 1889b: 396, 408).

SYNONYM: *Limnius mexicanus* Hinton, 1934: 199. — Type locality: Mexico.

latus Bug, 1973

Microcylloepus latus Bug, 1973: 124

TYPE LOCALITY: Peru.

DISTRIBUTION: Peru.

longipes (Grouvelle, 1889b)

Helmis longipes Grouvelle, 1889b: 408

TYPE LOCALITY: Brazil (Santa Catarina).

DISTRIBUTION: Brazil (Santa Catarina), Paraguay.

moapus fraxinus La Rivers, 1949a

Microcylloepus moapus fraxinus La Rivers, 1949a: 209

TYPE LOCALITY: USA (Nevada).

DISTRIBUTION: USA (Nevada).

NOTE: Might be a discrete species; synonymized with *moapus moapus* by Poole & Gentili (1996: 264) without presenting any evidence.

moapus moapus La Rivers, 1949a

Microcylloepus moapus La Rivers, 1949a: 205

TYPE LOCALITY: USA (Nevada).

DISTRIBUTION: USA (Nevada).

NOTE: Epithet name derives from the Piute adjective "moapa" (= muddy), latinized: moapus, -a, -um.

nomia Hinton, 1945c

Microcylloepus nomia Hinton, 1945c: 61

Type Locality: French Guiana.

Distribution: French Guiana.

Note: Epithet name is a noun in apposition (a nymph of Arcadia in Greek mythology).

obesus Hinton, 1940a

Microcylloepus obesus Hinton, 1940a: 305

Type Locality: Mexico.

Distribution: Costa Rica, Mexico.

ochus Hinton, 1940h

Microcylloepus ochus Hinton, 1940h: 245

Type Locality: Brazil (Santa Catarina).

Distribution: Brazil (Santa Catarina).

plaumanni Hinton, 1940d

Microcylloepus plaumanni Hinton, 1940d: 68

Type Locality: Brazil (Santa Catarina).

Distribution: Argentina, Brazil (Santa Catarina), Paraguay.

pumilus Bug, 1973

Microcylloepus pumilus Bug, 1973: 118

Type Locality: Brazil (São Paulo).

Distribution: Brazil (São Paulo).

pusillus (Le Conte, 1852)

Stenelmis pusilla Le Conte, 1852: 44

Type Locality: USA (New York).

Distribution: Canada (British Columbia, New Brunswick, Ontario, Quebec), USA (Alabama, Arkansas, Connecticut, District of Columbia, Florida, Georgia, Indiana, Kansas, Louisiana, Maine, Maryland, Mississippi, Montana, Nebraska, New Mexico, New York, Oklahoma, Pennsylvania, South Dakota, Tennessee, Texas, Utah, Virginia, West Virgina, Wisconsin, Wyoming).

Note: Incorrect original spelling: "*Stenelmis pusillus*"; incorrect subsequent spelling: "*Microcylloepus pusilla*" (Poole & Gentili 1996: 264).

Synonyms: *Elmis foveata* Leconte, 1874: 53. — Type locality: Unknown ("Two specimens without locality collected by Mr. Crotch" (Leconte 1874: 53)), the remaining elmids described together with *Elmis foveata* and collected by G.R. Crotch originate from California (USA) (2 spp.) and from British Columbia (Canada) (1 sp.). Note: Incorrect original spelling: "*Elmis foveatus*".

Helmis pusilla apta Musgrave, 1933: 56. — Type locality: USA (Virginia).

Helmis pusilla loedingi Musgrave, 1933: 56. — Type locality: USA (Mississippi). NOTE: Incorrect original spelling: "*lödingi*"; named after "Dr. H.P. Löding, of Mobile" (obviously a German family name); changed to *loedingi* by Brown (1983: 6) (ICZN 1999: Art. 32.5.2.1).

Helmis pusilla perdita Musgrave, 1933: 56. — Type locality: USA (Florida).

pustulatus Hinton, 1940d

Microcylloepus pustulatus Hinton, 1940d: 62

TYPE LOCALITY: Brazil (Amazonas).

DISTRIBUTION: Brazil (Amazonas).

similis (Horn, 1870)

Elmis similis Horn, 1870: 38

TYPE LOCALITY: USA (Arizona).

DISTRIBUTION: USA (Arizona, California, New Mexico).

sparsus Hinton, 1940h

Microcylloepus sparsus Hinton, 1940h: 237

TYPE LOCALITY: Brazil (Santa Catarina).

DISTRIBUTION: Brazil (Santa Catarina), Paraguay, Uruguay.

spinipes Hinton, 1940d

Microcylloepus spinipes Hinton, 1940d: 63

TYPE LOCALITY: Brazil (Santa Catarina).

DISTRIBUTION: Brazil (Santa Catarina), Paraguay.

steffani Bug, 1973

Microcylloepus steffani Bug, 1973: 123

TYPE LOCALITY: Peru.

DISTRIBUTION: Peru.

thermarum (Darlington, 1928)

Helmis thermarum Darlington, 1928

TYPE LOCALITY: USA (Nevada); in the original description the type locality was attributed to Oregon (p. 6: "Opal Mine 25 mi. So. Denio, Ore."), but in fact it belongs to Nevada (see Brown 1972a: 40, 1983: 7).

DISTRIBUTION: USA (Nevada).

troilus Hinton, 1940a

Microcylloepus troilus Hinton, 1940a: 314

Type Locality: Mexico.

Distribution: Mexico.

Microdinodes Grouvelle, 1906c

Microdinodes Grouvelle, 1906c: 324. — Type species: *Microdinodes quadrifasciatus* Grouvelle, 1906c. — Gender masculine.

Subgenus: ***Paramicrodinodes*** Delève, 1965d: 51. — Type species: *Microdinodes vaalensis* Delève, 1965d. — Gender masculine. Note: Incorrect subsequent spelling: "*Pammicrodinodes*" (http://www.organismnames.com).

balfouri Delève, 1967b

Microdinodes balfouri Delève, 1967b: 435

Type Locality: Zambia.

Distribution: Angola, Ghana, Liberia, Zambia.

basilewskyi (Janssens, 1962)

Aspidelmis basilewskyi Janssens, 1962: 433

Type Locality: Tanzania.

Distribution: Tanzania.

bimaculatus Delève, 1965d

Microdinodes bimaculatus Delève, 1965d: 11

Type Locality: Congo (DR).

Distribution: Angola, Congo (DR), Congo (R).

blickenstaffi Delève, 1973b

Microdinodes blickenstaffi Delève, 1973b: 307

Type Locality: Liberia.

Distribution: Liberia.

caelatus Delève, 1965d

Microdinodes caelatus Delève, 1965d: 30

Type Locality: Congo (DR).

Distribution: Congo (DR).

camerunensis Delève, 1963f

Microdinodes camerunensis Delève, 1963f: 816

Type Locality: Cameroon.

Distribution: Cameroon, Congo (R).

dahli Delève, 1963f

Microdinodes dahli Delève, 1963f: 819

Type Locality: Cameroon.

Distribution: Cameroon, Congo (R).

difficilis Grouvelle, 1911c

Microdinodes difficilis Grouvelle, 1911c: 268

Type Locality: Uganda.

Distribution: Congo (DR), Kenya, Uganda.

discedens Delève, 1965d

Microdinodes discedens Delève, 1965d: 15

Type Locality: Congo (DR).

Distribution: Congo (DR).

elegans Delève, 1965d

Microdinodes elegans Delève, 1965d: 39

Type Locality: Angola.

Distribution: Angola, Congo (DR).

gabonensis Delève, 1972

Microdinodes gabonensis Delève, 1972: 918

Type Locality: Gabon.

Distribution: Gabon.

garambanus Delève, 1963e

Microdinodes garambanus Delève, 1963e: 88

Type Locality: Congo (DR).

Distribution: Congo (DR).

guineensis Delève, 1965d

Microdinodes guineensis Delève, 1965d: 35

Type Locality: Guinea.

Distribution: Guinea.

illustris (Grouvelle, 1902a)

Helmis illustris Grouvelle, 1902a: 190

Type Locality: Congo (DR).

Distribution: Angola, Congo (DR), Congo (R), Ghana, Liberia.

imagineus Delève, 1965d

Microdinodes imagineus Delève, 1965d: 18

Type Locality: Congo (DR).

Distribution: Congo (DR).

Note: Incorrect subsequent spelling: "*imageus*" (Bertrand 1972: 505).

insolitus Delève, 1965d

Microdinodes insolitus Delève, 1965d: 48

Type Locality: Congo (DR).

Distribution: Congo (DR).

jeanneli Delève, 1946

Microdinodes jeanneli Delève, 1946: 327

Type Locality: Kenya.

Distribution: Kenya.

leleupi Delève, 1965d

Microdinodes leleupi Delève, 1965d: 28

Type Locality: Congo (DR).

Distribution: Congo (DR).

lituratus Delève, 1965d

Microdinodes lituratus Delève, 1965d: 49

Type Locality: Kenya.

Distribution: Congo (DR), Kenya.

marlieri Delève, 1965d

Microdinodes marlieri Delève, 1965d: 19

Type Locality: Congo (DR).

Distribution: Congo (DR), Rwanda, Uganda.

melaenus Grouvelle, 1906c

Microdinodes melaenus Grouvelle, 1906c: 325

Type Locality: Tanzania.

Distribution: Kenya, Tanzania.

multimaculatus Delève, 1965d

Microdinodes multimaculatus Delève, 1965d: 37

Type Locality: Congo (DR).

Distribution: Angola, Congo (DR).

nigrolineatus Delève, 1937a

Microdinodes nigrolineatus Delève, 1937a: 156

Type Locality: Congo (DR).

Distribution: Angola, Congo (DR), Congo (R), Liberia.

obscurus Delève, 1965d

Microdinodes obscurus Delève, 1965d: 46

Type Locality: Congo (DR).

Distribution: Congo (DR).

octoguttatus Delève, 1967a

Microdinodes octoguttatus Delève, 1967a: 328

Type Locality: Congo (R).

Distribution: Congo (R).

ornatus Grouvelle, 1911c

Microdinodes ornatus Grouvelle, 1911c: 267

Type Locality: Uganda.

Distribution: Congo (DR), Uganda.

Synonym: *Microdinodes vageguttatus* Grouvelle, 1911c: 270. — Type locality: Uganda. Note: Incorrect subsequent spelling: "*vagepunctatus*" (Jeannel 1950: 170); in the original description, the habitus of *M. vageguttatus* is shown in pl. 1: fig. 3 (the caption for fig. 3 erroneously refers to *Protelmis limnioides* [Protelmidae]).

parallelus Delève, 1946

Microdinodes parallelus Delève, 1946: 328

Type Locality: Kenya.

Distribution: Kenya.

pilistriatus Delève, 1965d

Microdinodes pilistriatus Delève, 1965d: 42

Type Locality: South Africa.

Distribution: South Africa.

quadrifasciatus Grouvelle, 1906c

Microdinodes quadrifasciatus Grouvelle, 1906c: 324

Type Locality: Tanzania.

Distribution: Kenya, Tanzania.

quadrisignatus Grouvelle, 1911c

Microdinodes quadrisignatus Grouvelle, 1911c: 271

Type Locality: Uganda.

Distribution: Congo (DR), Kenya, Tanzania, Uganda.

sexualis Delève, 1965d

Microdinodes sexualis Delève, 1965d: 45

Type Locality: Congo (DR).

Distribution: Congo (DR), Uganda.

similis Delève, 1937a

Microdinodes similis Delève, 1937a: 158

Type Locality: Congo (DR).

Distribution: Congo (DR), Congo (R).

simoni (Grouvelle, 1895)

Stenelmis simoni Grouvelle, 1895: 168

Type Locality: South Africa.

Distribution: Angola, Congo (DR), Kenya, South Africa.

Synonym: *Microdinodes immaculatus* Delève, 1937a: 158. — Type locality: Congo (DR).

tarsalis Delève, 1942

Microdinodes tarsalis Delève, 1942: 5

Type Locality: Congo (DR).

Distribution: Congo (DR), Congo (R).

tibialis Delève, 1938

Microdinodes tibialis Delève, 1938: 368

Type Locality: Congo (DR).

Distribution: Congo (DR), Congo (R), Rwanda.

transvaalicus (Grouvelle, 1895)

Stenelmis transvaalica Grouvelle, 1895: 167

Type Locality: South Africa.

Distribution: Congo (DR), South Africa.

Synonym: *Microdinodes vittatus* Delève, 1938: 366. — Type locality: Congo (DR).

troilus Hinton, 1940g

Microdinodes troilus Hinton, 1940g: 301

Type Locality: Ethiopia.

Distribution: Ethiopia.

vaalensis Delève, 1965d

Microdinodes vaalensis Delève, 1965d: 51 (subgenus *Paramicrodinodes*)
Type Locality: South Africa.
Distribution: South Africa.

venustus Delève, 1965d

Microdinodes venustus Delève, 1965d: 6
Type Locality: Congo (DR).
Distribution: Congo (DR), Congo (R).

villiersi Delève, 1967a

Microdinodes villiersi Delève, 1967a: 326
Type Locality: Congo (R).
Distribution: Congo (R).

zambesinus (Brancsik, 1914)

Riolus zambesinus Brancsik, 1914: 61
Type Locality: Mozambique.
Distribution: Mozambique.
Note: Redescribed by Delève (1967b: 433).

Microlara Jäch, 1993

Microlara Jäch, 1993: 15. — Type species: *Microlara mahensis* Jäch, 1993. — Gender feminine.

mahensis Jäch, 1993

Microlara mahensis Jäch, 1993: 16
Type Locality: Seychelles.
Distribution: Seychelles.

Narpus Casey, 1893

Narpus Casey, 1893: 582. — Type species: *Narpus angustus* Casey, 1893. — Gender masculine.

angustus Casey, 1893

Narpus angustus Casey, 1893: 583
Type Locality: USA (California).
Distribution: USA (California).

arizonicus (Brown, 1930b)

Helmis arizonica Brown, 1930b: 90

Type Locality: USA (Arizona).

Distribution: USA (Arizona).

Note: Incorrect subsequent spellings: "*Helmis arizonicus*" (Brown 1933: 47), "*Narpus arizonica*" (Poole & Gentili 1996: 264).

concolor (Leconte, 1881)

Elmis concolor Leconte, 1881: 72

Type Locality: USA (New Mexico).

Distribution: Canada (Alberta, British Columbia), USA (Arizona, California, Nevada, New Mexico, Oregon, Utah, Washington, Wyoming).

Synomym: *Helmis soluta* Brown, 1933: 46. — Type locality: Canada (British Columbia). Note: Incorrect original spelling: "*solutus*".

Neblinagena Spangler, 1985a

Neblinagena Spangler, 1985a: 539. — Type species: *Neblinagena prima* Spangler, 1985a. — Gender feminine.

doylei Kodada & Jäch, 1999

Neblinagena doylei Kodada & Jäch, 1999: 27

Type Locality: Venezuela.

Distribution: Venezuela.

prima Spangler, 1985a

Neblinagena prima Spangler, 1985a: 541

Type Locality: Venezuela.

Distribution: Venezuela.

Neocylloepus Brown, 1970a

Neocylloepus Brown, 1970a: 2. — Type species: *Elmis sculptipennis* Sharp, 1882. — Gender masculine.

arringtoni Brown, 1970a

Neocylloepus arringtoni Brown, 1970a: 19

Type Locality: Mexico.

Distribution: Mexico.

boeseli Brown, 1970a

Neocylloepus boeseli Brown, 1970a: 15

TYPE LOCALITY: USA (Texas).

DISTRIBUTION: Mexico, USA (Texas).

championi (Sharp, 1882)

Elmis championi Sharp, 1882: 134

TYPE LOCALITY: Panama.

DISTRIBUTION: Panama.

chaparensis Manzo & Moya, 2010

Neocylloepus chaparensis Manzo & Moya, 2010: 130

TYPE LOCALITY: Bolivia.

DISTRIBUTION: Bolivia.

hintoni Brown, 1970a

Neocylloepus hintoni Brown, 1970a: 12

TYPE LOCALITY: Mexico.

DISTRIBUTION: Mexico.

petersoni Brown, 1970a

Neocylloepus petersoni Brown, 1970a: 18

TYPE LOCALITY: Mexico.

DISTRIBUTION: Mexico.

sandersoni Brown, 1970a

Neocylloepus sandersoni Brown, 1970a: 17

TYPE LOCALITY: Costa Rica.

DISTRIBUTION: Colombia, Costa Rica, Panama.

sculptipennis (Sharp, 1882)

Elmis sculptipennis Sharp, 1882: 135

TYPE LOCALITY: Guatemala.

DISTRIBUTION: Belize, Costa Rica, Guatemala, Honduras, Mexico, Nicaragua.

NOTE: Incorrect subsequent spelling: "*scultipennis*" (Manzo 2013: 211).

Neoelmis Musgrave, 1935

Neoelmis Musgrave, 1935: 34. — Type species: *Neoelmis gracilis* Musgrave, 1935. — Gender feminine.

NOTE: Incorrect subsequent spelling: "*Neoelmus*" (Peck et al. 2003: 14).

abdominalis Hinton, 1939d

Neoelmis abdominalis Hinton, 1939d: 231

Type Locality: Colombia.

Distribution: Colombia.

alcine Hinton, 1972b

Neoelmis alcine Hinton, 1972b: 124

Type Locality: Brazil (Paraná).

Distribution: Brazil (Paraná).

ampla Hinton, 1940e

Neoelmis ampla Hinton, 1940e: 144

Type Locality: Brazil (Rondônia); erroneously, Hinton (1940e: 144) attributed the type locality to Mato Grosso ("Matto [sic] Grosso, Rio Giparana, 4.ix.1937 (H.E. Hinton)").

Distribution: Brazil (Rondônia); erroneously reported from Mato Grosso by Segura et al. (2013: 33).

anytis Hinton, 1972b

Neoelmis anytis Hinton, 1972b: 133

Type Locality: Ecuador.

Distribution: Ecuador.

apicalis angusta Hinton, 1939d

Neoelmis apicalis angusta Hinton, 1939d: 233

Type Locality: Peru.

Distribution: Bolivia, Peru.

Note: Probably a discrete species.

apicalis apicalis (Sharp, 1882)

Elmis apicalis Sharp, 1882: 136

Type Locality: Guatemala.

Distribution: Costa Rica, Guatemala, Mexico.

aragua Hinton, 1972c

Neoelmis aragua Hinton, 1972c: 143

Type Locality: Venezuela.

Distribution: Venezuela.

argentinensis Manzo & Archangelsky, 2012

Neoelmis argentinensis Manzo & Archangelsky, 2012: 272

TYPE LOCALITY: Argentina.

DISTRIBUTION: Argentina.

aspera Hinton, 1940a

Neoelmis aspera Hinton, 1940a: 330

TYPE LOCALITY: Mexico.

DISTRIBUTION: Mexico, Nicaragua, Panama.

atys Hinton, 1972b

Neoelmis atys Hinton, 1972b: 125

TYPE LOCALITY: Brazil (Rio Grande do Sul).

DISTRIBUTION: Brazil (Rio Grande do Sul).

azteca Hinton, 1940a

Neoelmis azteca Hinton, 1940a: 327

TYPE LOCALITY: Mexico.

DISTRIBUTION: Mexico.

caesa (Leconte, 1874)

Elmis caesa Leconte, 1874: 53

TYPE LOCALITY: USA (Texas).

DISTRIBUTION: USA (Oklahoma, Texas).

NOTE: Incorrect original spelling: "*Elmis caesus*"; incorrect subsequent spelling: "*Helmis cæsa*" (Darlington 1936: 79).

ceto Hinton, 1972c

Neoelmis ceto Hinton, 1972c: 137

TYPE LOCALITY: Venezuela.

DISTRIBUTION: Venezuela.

crino Hinton, 1972c

Neoelmis crino Hinton, 1972c: 135

TYPE LOCALITY: Venezuela.

DISTRIBUTION: Venezuela.

fossa Hinton, 1940e

Neoelmis fossa Hinton, 1940e: 145

TYPE LOCALITY: Brazil (Amazonas).

DISTRIBUTION: Brazil (Amazonas).

giga Hinton, 1939e

Neoelmis giga Hinton, 1939e: 39

TYPE LOCALITY: Brazil (Santa Catarina).

DISTRIBUTION: Brazil (Santa Catarina).

gracilis Musgrave, 1935

Neoelmis gracilis Musgrave, 1935: 35

TYPE LOCALITY: Puerto Rico.

DISTRIBUTION: Puerto Rico.

grossa Hinton, 1939d

Neoelmis grossa Hinton, 1939d: 228

TYPE LOCALITY: Bolivia.

DISTRIBUTION: Bolivia.

grossepunctata Delève, 1968a

Neoelmis grossepunctata Delève, 1968a: 236

TYPE LOCALITY: Ecuador.

DISTRIBUTION: Ecuador.

limosa (Grouvelle, 1908)

Helmis limosa Grouvelle, 1908: 184

TYPE LOCALITY: French Guiana.

DISTRIBUTION: Brazil (Amazonas, Pará, Rondônia), French Guiana.

lobata Hinton, 1939e

Neoelmis lobata Hinton, 1939e: 40

TYPE LOCALITY: Brazil (Santa Catarina).

DISTRIBUTION: Brazil (Santa Catarina).

longula Hinton, 1936c

Neoelmis longula Hinton, 1936c: 426

TYPE LOCALITY: Mexico.

DISTRIBUTION: Mexico.

NOTE: Incorrect original spelling: "*longulus*".

maculata Hinton, 1940e

Neoelmis maculata Hinton, 1940e: 140

TYPE LOCALITY: Brazil (Amazonas).

DISTRIBUTION: Brazil (Amazonas), Paraguay.

marmorata Hinton, 1940e

Neoelmis marmorata Hinton, 1940e: 143

Type Locality: Brazil (Rondônia); erroneously, Hinton (1940e: 144) attributed the type locality to Mato Grosso ("Matto [sic] Grosso, Rio Candeïa, 30.viii.1937 (H.E. Hinton)").

Distribution: Brazil (Rondônia); erroneously reported from Mato Grosso by Segura et al. (2013: 35).

maro Hinton, 1972b

Neoelmis maro Hinton, 1972b: 120

Type Locality: Brazil (Paraná).

Distribution: Brazil (Paraná).

mila Hinton, 1972b

Neoelmis mila Hinton, 1972b: 131

Type Locality: Brazil (Rio Grande do Sul).

Distribution: Brazil (Rio Grande do Sul).

minima (Darlington, 1927)

Helmis minima Darlington, 1927: 93

Type Locality: Cuba.

Distribution: Cuba, Haiti, Jamaica.

morador Hinton, 1972c

Neoelmis morador Hinton, 1972c: 138

Type Locality: Venezuela.

Distribution: Venezuela.

mormo Hinton, 1972b

Neoelmis mormo Hinton, 1972b: 127

Type Locality: Brazil (Rio Grande do Sul).

Distribution: Brazil (Paraná, Rio Grande do Sul).

musgravei Hinton, 1940e

Neoelmis musgravei Hinton, 1940e: 150

Type Locality: Brazil (Santa Catarina).

Distribution: Brazil (Santa Catarina).

nana Hinton, 1940e

Neoelmis nana Hinton, 1940e: 147

Type Locality: Brazil (Santa Catarina).

Distribution: Brazil (Santa Catarina).

nelo Hinton, 1972b

Neoelmis nelo Hinton, 1972b: 130

Type Locality: Brazil (Santa Catarina).

Distribution: Brazil (Santa Catarina), Paraguay.

nicon Hinton, 1972b

Neoelmis nicon Hinton, 1972b: 128

Type Locality: Brazil (Santa Catarina).

Distribution: Brazil (Paraná, Rio Grande do Sul, Santa Catarina).

olenus Hinton, 1972c

Neoelmis olenus Hinton, 1972c: 143

Type Locality: Venezuela.

Distribution: Venezuela.

Note: Epithet name is a noun in apposition referring to several figures in Greek mythology, e.g. the son of Hephaestus.

opis Hinton, 1972b

Neoelmis opis Hinton, 1972b: 134

Type Locality: Bolivia.

Distribution: Bolivia, Brazil (Amazonas), Ecuador, Paraguay.

plaumanni Hinton, 1940e

Neoelmis plaumanni Hinton, 1940e: 138

Type Locality: Brazil (Santa Catarina).

Distribution: Brazil (Santa Catarina).

porrecta Delève, 1968a

Neoelmis porrecta Delève, 1968a: 241

Type Locality: Ecuador.

Distribution: Ecuador.

prosternalis Hinton, 1939e

Neoelmis prosternalis Hinton, 1939e: 38

Type Locality: Brazil (Santa Catarina).

Distribution: Brazil (Santa Catarina).

pusio Hinton, 1971d

Neoelmis pusio Hinton, 1971d: 256

Type Locality: Trinidad and Tobago (Tobago).

Distribution: Trinidad and Tobago (Trinidad, Tobago).

reichardti Hinton, 1972b

Neoelmis reichardti Hinton, 1972b: 118

Type Locality: Brazil (Santa Catarina).

Distribution: Brazil (Santa Catarina).

resa Hinton, 1972c

Neoelmis resa Hinton, 1972c: 140

Type Locality: Venezuela.

Distribution: Venezuela.

saon Hinton, 1972c

Neoelmis saon Hinton, 1972c: 142

Type Locality: Venezuela.

Distribution: Venezuela.

scissicollis (Germain, 1892)

Elmis scissicollis Germain, 1892: 246

Type Locality: Chile.

Distribution: Chile.

simoni (Grouvelle, 1889c)

Helmis simoni Grouvelle, 1889c: 164

Type Locality: Venezuela.

Distribution: Venezuela.

sketi Spangler, 1996b

Neoelmis sketi Spangler, 1996b: 246

Type Locality: Ecuador.

Distribution: Ecuador.

sul Hinton, 1972b

Neoelmis sul Hinton, 1972b: 123

Type Locality: Brazil (Goiás).

Distribution: Brazil (Goiás).

thoracica (Grouvelle, 1896a)

Helmis thoracica Grouvelle, 1896a: 48

TYPE LOCALITY: Bolivia.

DISTRIBUTION: Bolivia.

thyas Hinton, 1972b

Neoelmis thyas Hinton, 1972b: 122

TYPE LOCALITY: Brazil (Rondônia).

DISTRIBUTION: Brazil (Rondônia).

NOTE: Seems to be intermediate between *Neoelmis* Musgrave, 1935 and *Elachistelmis* Maier, 2012.

tibialis Delève, 1968a

Neoelmis tibialis Delève, 1968a: 239

TYPE LOCALITY: Ecuador.

DISTRIBUTION: Ecuador.

tocuyito Hinton, 1972c

Neoelmis tocuyito Hinton, 1972c: 137

TYPE LOCALITY: Venezuela.

DISTRIBUTION: Venezuela.

Neolimnius Hinton, 1939e

Neolimnius Hinton, 1939e: 41 — Type species: *Neolimnius palpalis* Hinton, 1939e. — Gender masculine.

palpalis Hinton, 1939e

Neolimnius palpalis Hinton, 1939e: 43

TYPE LOCALITY: Brazil (Amazonas); erroneously, Hinton (1939e: 45) attributed the type locality to the State of Pará.

DISTRIBUTION: Brazil (Amazonas), Colombia, French Guiana.

Neoriohelmis Nomura, 1958b

Neoriohelmis Nomura, 1958b: 45. — Type species: *Neoriohelmis kurosawai* Nomura, 1958b. — Gender feminine.

NOTE: This genus is obviously a junior synonym of *Narpus* Casey, 1893.

kurosawai Nomura, 1958b

Neoriohelmis kurosawai Nomura, 1958b: 46

Type Locality: Japan (Honshu).

Distribution: Japan (Honshu).

kuwatai Satô, 1963b

Neoriohelmis kuwatai Satô, 1963b: 132

Type Locality: Japan (Shikoku).

Distribution: Japan (Shikoku).

Nesonychus Jäch & Boukal, 1997a

Nesonychus Jäch & Boukal, 1997a: 214. — Type species: *Nesonychus kodadai* Jäch & Boukal, 1997a. — Gender masculine.

gibbulus Jäch & Boukal, 1997a

Nesonychus gibbulus Jäch & Boukal, 1997a: 217

Type Locality: Indonesia (Sumatra).

Distribution: Indonesia (Sumatra).

kodadai Jäch & Boukal, 1997a

Nesonychus kodadai Jäch & Boukal, 1997a: 216

Type Locality: Malaysia (Sarawak).

Distribution: Malaysia (Sabah, Sarawak).

Nomuraelmis Satô, 1964a

Nomuraelmis Satô, 1964a: 11. — Type species: *Nomuraelmis amamiensis* Satô, 1964a. — Gender feminine.

amamiensis Satô, 1964a

Nomuraelmis amamiensis Satô, 1964a: 12

Type Locality: Japan (Ryukyu Islands).

Distribution: Japan (Ryukyu Islands).

Notelmis Hinton, 1941b

Notelmis Hinton, 1941b: 65. — Type species: *Elmis nodipes* Sharp, 1882. — Gender feminine.

Note: This genus is very similar to *Onychelmis* Hinton, 1941b.

bifoveolata Delève, 1968a

Notelmis bifoveolata Delève, 1968a: 220

Type Locality: Ecuador.

Distribution: Ecuador.

nodipes (Sharp, 1882)

Elmis nodipes Sharp, 1882: 138

Type Locality: Panama.

Distribution: Costa Rica, Panama.

Notriolus Carter & Zeck, 1929

Notriolus Carter & Zeck, 1929: 63. — Type species: *Elmis quadriplagiata* Carter, 1926b. — Gender masculine.

allynensis (Carter, 1926b)

Helmis allynensis Carter, 1926b: 61

Type Locality: Australia (New South Wales).

Distribution: Australia (New South Wales).

barretti (Carter, 1926a)

Helmis barretti Carter, 1926a: 506

Type Locality: Australia (New South Wales).

Distribution: Australia (New South Wales).

Synonym: *Helmis barretti basalis* Carter, 1926a: 507. — Type locality: Australia (New South Wales). Note: Originally described as "var. *basalis*"; might be a good species.

davidsoni Carter & Zeck, 1936

Notriolus davidsoni Carter & Zeck, 1936: 156

Type Locality: Australia (New South Wales).

Distribution: Australia (New South Wales).

dorrigoensis Carter & Zeck, 1929

Notriolus dorrigoensis Carter & Zeck, 1929: 64

Type Locality: Australia (New South Wales).

Distribution: Australia (New South Wales).

galstonius Carter & Zeck, 1929

Notriolus galstonius Carter & Zeck, 1929: 65

Type Locality: Australia (New South Wales).

Distribution: Australia (New South Wales).

Note: Partly misspelled ("*galstonensis*") in original description (p. 70).

humeralis Carter & Zeck, 1929

Notriolus humeralis Carter & Zeck, 1929: 65

TYPE LOCALITY: Australia (New South Wales).
DISTRIBUTION: Australia (New South Wales).

maculatus (Carter, 1926a)

Helmis maculata Carter, 1926a: 507
TYPE LOCALITY: Australia (New South Wales).
DISTRIBUTION: Australia (New South Wales).

minor Carter & Zeck, 1932

Notriolus minor Carter & Zeck, 1932: 204
TYPE LOCALITY: Australia (New South Wales).
DISTRIBUTION: Australia (New South Wales).

minutus Carter & Zeck, 1938

Notriolus minutus Carter & Zeck, 1938: 171
TYPE LOCALITY: Australia (Queensland).
DISTRIBUTION: Australia (Queensland).

quadriplagiatus (Carter, 1926b)

Helmis quadriplagiata Carter, 1926b: 63
TYPE LOCALITY: Australia (Victoria).
DISTRIBUTION: Australia (Tasmania, Victoria).

setosus Carter & Zeck, 1936

Notriolus setosus Carter & Zeck, 1936: 156
TYPE LOCALITY: Australia (New South Wales).
DISTRIBUTION: Australia (New South Wales).

simsoni (Grouvelle, 1896a)

Helmis simsoni Grouvelle, 1896a: 47
TYPE LOCALITY: Australia (Tasmania).
DISTRIBUTION: Australia (Tasmania).

subplanatus Carter & Zeck, 1929

Notriolus subplanatus Carter & Zeck, 1929: 66
TYPE LOCALITY: Australia (Queensland).
DISTRIBUTION: Australia (Queensland).

taylori Carter & Zeck, 1936

Notriolus taylori Carter & Zeck, 1936: 157

Type Locality: Australia (Queensland).
Distribution: Australia (Queensland).

tropicus Carter & Zeck, 1938

Notriolus tropicus Carter & Zeck, 1938: 171
Type Locality: Australia (Queensland).
Distribution: Australia (Queensland).

victoriae Carter & Zeck, 1929

Notriolus victoriae Carter & Zeck, 1929: 66
Type Locality: Australia (Victoria).
Distribution: Australia (New South Wales?, Victoria).

Ohiya Jäch, 1982b

Ohiya Jäch, 1982b: 96. — Type species: *Ohiya carinata* Jäch, 1982b. — Gender feminine.

carinata Jäch, 1982b

Ohiya carinata Jäch, 1982b: 96
Type Locality: Sri Lanka.
Distribution: Sri Lanka.

Okalia Kodada & Čiampor, 2003

Okalia Kodada & Čiampor, 2003: 784. — Type species: *Okalia globosa* Kodada & Čiampor, 2003. — Gender feminine.

globosa Kodada & Čiampor, 2003

Okalia globosa Kodada & Čiampor, 2003: 791
Type Locality: Malaysia (Pahang).
Distribution: Malaysia (Pahang).

Omotonus Delève, 1963c

Omotonus Delève, 1963c: 433. — Type species: *Potamophilinus notabilis* Grouvelle, 1898b. — Gender masculine.

angolensis Delève, 1963c

Omotonus angolensis Delève, 1963c: 437
Type Locality: Angola.
Distribution: Angola, Liberia.

bertrandi Delève, 1963c

Omotonus bertrandi Delève, 1963c: 439

TYPE LOCALITY: Angola.

DISTRIBUTION: Angola.

browni Więźlak, 1987b

Omotonus browni Więźlak, 1987b: 444

TYPE LOCALITY: Cameroon.

DISTRIBUTION: Cameroon.

kwangolensis Delève, 1963c

Omotonus kwangolensis Delève, 1963c: 436

TYPE LOCALITY: Congo (DR).

DISTRIBUTION: Congo (DR), Congo (R).

notabilis (Grouvelle, 1898b)

Potamophilinus notabilis Grouvelle, 1898b: 112

TYPE LOCALITY: Congo (DR).

DISTRIBUTION: Congo (DR), Congo (R), Zambia.

spinicaudus (Hinton, 1935)

Potamophilinus spinicaudus Hinton, 1935: 175

TYPE LOCALITY: Sierra Leone.

DISTRIBUTION: Congo (DR), Congo (R), Gabon, Sierra Leone.

Onychelmis Hinton, 1941b

Onychelmis Hinton, 1941b: 66. — Type species: *Elmis longicollis* Sharp, 1882. — Gender feminine.

NOTE: This genus is very similar to *Notelmis* Hinton, 1941b.

leleupi Delève, 1968a

Onychelmis leleupi Delève, 1968a: 217

TYPE LOCALITY: Ecuador.

DISTRIBUTION: Ecuador.

longicollis (Sharp, 1882)

Elmis longicollis Sharp, 1882: 138

TYPE LOCALITY: Panama.

DISTRIBUTION: Colombia, Costa Rica, Guatemala, Panama.

whiteheadi Spangler & Santiago, 1991

Onychelmis whiteheadi Spangler & Santiago, 1991: 495

Type Locality: Colombia.

Distribution: Colombia.

Oolimnius Hinton, 1939e

Oolimnius Hinton, 1939e: 36. — Type species: *Oolimnius salti* Hinton, 1939e. — Gender masculine.

salti Hinton, 1939e

Oolimnius salti Hinton, 1939e: 37

Type Locality: Brazil (Santa Catarina).

Distribution: Brazil (Goiás, Santa Catarina).

Optioservus Sanderson, 1953a

Optioservus Sanderson, 1953a: 155. — Type species: *Limnius trivittatus* Brown, 1930b. — Gender masculine.

Note: Originally described in a key; formal description published by Sanderson (1954: 8); might be a synonym of *Heterlimnius* Hinton, 1935.

browni White, 1978

Optioservus browni White, 1978: 64

Type Locality: USA (Arkansas).

Distribution: USA (Arkansas).

canus Chandler, 1954

Optioservus canus Chandler, 1954: 130

Type Locality: USA (California).

Distribution: USA (California).

castanipennis (Fall, 1925)

Helmis castanipennis Fall, 1925: 177

Type Locality: USA (Wyoming).

Distribution: USA (Colorado, New Mexico, Utah, Wyoming).

divergens (Leconte, 1874)

Elmis divergens Leconte, 1874: 52

Type Locality: USA (California).

Distribution: Canada (Alberta, British Columbia, Saskatchewan), USA (Arizona, California, Colorado, Nevada, New Mexico, South Dakota, Utah, Wyoming).

Synonym: *Elmis pecosensis* Fall, 1907: 226. — Type locality: USA (New Mexico).

fastiditus (Leconte, 1850)

Limnius fastiditus Leconte, 1850: 217

Type Locality: Canada (Ontario).

Distribution: Canada (Alberta, British Columbia, Manitoba, New Brunswick, Newfoundland, Northwest Territories, Nova Scotia, Ontario, Prince Edward Island, Quebec, Saskatchewan), USA (Minnesota, Montana, North Dakota, Wisconsin).

gapyeongensis Jung, Kamite & Bae, 2011

Optioservus gapyeongensis Jung, Kamite & Bae, 2011: 179

Type Locality: Korea (R).

Distribution: China (Liaoning), (Korea (DPR)), Korea (R), Russia (Far East).

hagai Nomura, 1958b

Optioservus hagai Nomura, 1958b: 56

Type Locality: Japan (Kyushu).

Distribution: Japan (Kyushu).

heteroclitus White, 1978

Optioservus heteroclitus White, 1978: 68

Type Locality: USA (California).

Distribution: USA (California).

Note: Might be a junior synonym of *Optioservus divergens* (Leconte, 1874).

immunis (Fall, 1925)

Helmis immunis Fall, 1925: 178

Type Locality: USA (Connecticut).

Distribution: USA (Connecticut, Georgia, New Jersey, North Carolina, Pennsylvania, South Carolina, Tennessee).

Synonym: *Limnius cryophilus* Musgrave, 1932: 79. — Type locality: USA (Tennessee).

maculatus Nomura, 1958b

Optioservus maculatus Nomura, 1958b: 50

Type Locality: Japan (Honshu).

Distribution: Japan (Honshu).

nitidus Nomura, 1958b

Optioservus nitidus Nomura, 1958b: 53

TYPE LOCALITY: Japan (Honshu).

DISTRIBUTION: Japan (Honshu).

ovalis (Leconte, 1863)

Limnius ovalis Leconte, 1863: 74

TYPE LOCALITY: USA (Pennsylvania).

DISTRIBUTION: Canada (Labrador, New Brunswick, Nova Scotia, Ontario, Prince Edward Island, Quebec), USA (Alabama, Connecticut, Maryland, Massachusetts, Mississippi, New York, Ohio, Pennsylvania, Vermont, Virginia, West Virginia).

SYNONYM: *Helmis ampliata* Fall, 1925: 179. — Type locality: USA (Massachusetts). NOTE: Incorrect original spelling: "*ampliatus*".

phaeus White, 1978

Optioservus phaeus White, 1978: 70

TYPE LOCALITY: USA (Kansas).

DISTRIBUTION: USA (Kansas).

quadrimaculatus (Horn, 1870)

Elmis quadrimaculata Horn, 1870: 37

TYPE LOCALITY: USA (California).

DISTRIBUTION: Canada (Alberta, British Columbia), USA (California, Colorado, Idaho, Montana, Nevada, Oregon, Utah, Washington).

NOTE: Incorrect original spelling: "*Elmis quadrimaculatus*"; incorrect subsequent spellings: "*4-maculatus*" (Crotch 1873: 52, 53, Leconte 1874: 53), "*Optioservus quadrimaculata*" (Poole & Gentili 1996: 265).

rugulosus Nomura, 1958b

Optioservus rugulosus Nomura, 1958b: 53

TYPE LOCALITY: Japan (Kyushu).

DISTRIBUTION: Japan (Honshu, Kyushu).

sandersoni Collier, 1972

Optioservus sandersoni Collier, 1972: 18

TYPE LOCALITY: USA (Arkansas).

DISTRIBUTION: USA (Arkansas, Kansas, Missouri, Oklahoma).

SYNONYM: *Optioservus ozarkensis* Collier, 1972: 17. — Type locality: USA (Missouri).

seriatus (Leconte, 1874)

Elmis seriata Leconte, 1874: 52

TYPE LOCALITY: USA (California).

DISTRIBUTION: Canada (Alberta, British Columbia), USA (California, Colorado, Idaho, Oregon, Utah, Wyoming).

NOTE: Incorrect original spelling: "*Elmis seriatus*"; incorrect subsequent spelling: "*Optioservus seriata*" (Poole & Gentili 1996: 265).

trivittatus (Brown, 1930b)

Limnius trivittatus Brown, 1930b: 91

TYPE LOCALITY: Canada (Quebec).

DISTRIBUTION: Canada (New Brunswick, Newfoundland, Nova Scotia, Ontario, Quebec), USA (Indiana, Kentucky, Maryland, Minnesota, Mississippi, Missouri, New Jersey, New York, Ohio, Pennsylvania, South Carolina, Tennessee, Vermont, Virginia, West Virginia, Wisconsin).

variabilis Nomura, 1958b

Optioservus variabilis Nomura, 1958b: 51

TYPE LOCALITY: Japan (Honshu).

DISTRIBUTION: Japan (Honshu).

Ordobrevia Sanderson, 1953a

Ordobrevia Sanderson, 1953a: 159. — Type species: *Stenelmis nubifer* Fall, 1901. — Gender feminine.

NOTE: This genus is poorly defined (see Jäch 1984d: 284–285), very similar to *Stenelmis* Dufour, 1835.

amamiensis amamiensis (Nomura, 1957a)

Stenelmis amamiensis Nomura, 1957a: 4

TYPE LOCALITY: Japan (Ryukyu Islands).

DISTRIBUTION: Japan (Ryukyu Islands).

amamiensis okinawana Nomura, 1959

Ordobrevia amamiensis okinawana Nomura, 1959: 34

TYPE LOCALITY: Japan (Ryukyu Islands).

DISTRIBUTION: Japan (Ryukyu Islands).

communis Delève, 1968c

Ordobrevia communis Delève, 1968c: 168

TYPE LOCALITY: Vietnam.

DISTRIBUTION: Vietnam.

constricta Delève, 1968c

Ordobrevia constricta Delève, 1968c: 167

TYPE LOCALITY: Vietnam.

DISTRIBUTION: Vietnam.

flavolineata Delève, 1973a

Ordobrevia fletcheri flavolineata Delève, 1973a: 11

TYPE LOCALITY: Sri Lanka.

DISTRIBUTION: Sri Lanka.

fletcheri (Champion, 1923)

Stenelmis fletcheri Champion, 1923: 166

TYPE LOCALITY: Sri Lanka.

DISTRIBUTION: Sri Lanka.

foveicollis (Schönfeldt, 1888)

Stenelmis foveicollis Schönfeldt, 1888: 193

TYPE LOCALITY: Japan (Honshu).

DISTRIBUTION: Japan (Honshu, Kyushu, Shikoku).

SYNONYMS: *Stenelmis flavovittata* Kôno, 1934: 126. — Type locality: Japan (Honshu). NOTE: Incorrect original spelling: "*flavovittatus*".

Stenelmis freyi Bollow, 1941: 21. — Type locality: Japan (Honshu).

Stenelmis japonica Janssens, 1956: 2. — Type locality: Japan (Honshu).

gotoi Nomura, 1959

Ordobrevia gotoi Nomura, 1959: 33

TYPE LOCALITY: Japan (Honshu).

DISTRIBUTION: Japan (Honshu).

longicollis (Pic, 1923)

Stenelmis longicollis Pic, 1923: 4

TYPE LOCALITY: Vietnam.

DISTRIBUTION: Vietnam.

maculata (Nomura, 1957b)

Stenelmis maculata Nomura, 1957b: 43

TYPE LOCALITY: Japan (Honshu).

DISTRIBUTION: Japan (Honshu, Shikoku, Yakushima).

nubifer (Fall, 1901)

Stenelmis nubifer Fall, 1901: 238

TYPE LOCALITY: USA (California).

DISTRIBUTION: Canada (British Columbia), USA (California, Oregon, Washington).

NOTE: Incorrect subsequent spelling: "*nubifera*" (Zaitzev 1908: 300, 1910: 22) (see ICZN 1999: Art. 31.2.2).

reflexicollis (Bollow, 1940b)

Stenelmis reflexicollis Bollow, 1940b: 16

TYPE LOCALITY: Myanmar.

DISTRIBUTION: Myanmar.

Orientelmis Shepard, 1998

Orientelmis Shepard, 1998: 289. — Type species: *Orientelmis sinensis* Shepard, 1998. — Gender feminine.

parvula (Nomura & Baba, 1961)

Cleptelmis parvula Nomura & Baba, 1961: 4

TYPE LOCALITY: Japan (Honshu).

DISTRIBUTION: Japan (Honshu, Kyushu).

sinensis Shepard, 1998

Orientelmis sinensis Shepard, 1998: 291

TYPE LOCALITY: China (Jiangxi).

DISTRIBUTION: China (Guangxi, Hunan, Jiangxi).

Oulimnius Gozis, 1886

Oulimnius Gozis, 1886: 9. — Type species: *Limnius tuberculatus* Müller, 1806a. — Gender masculine.

NOTE: Incorrect subsequent spelling: "*Ulimnius*" (Grouvelle 1896e: 27).

aegyptiacus (Kuwert, 1890)

Limnius aegyptiacus Kuwert, 1890: 44

TYPE LOCALITY: Egypt (details unknown).

DISTRIBUTION: Egypt, Morocco? (Berthélemy 1979: 74), Spain? (Berthélemy 1979: 74).

SYNONYM: *Limnius aegyptiacus lineatus* Kuwert, 1890: 20. — Type locality: Egypt (Sinai).
NOTE: Originally described as a variety of *Limnius aegyptiacus*.

bertrandi Berthélemy, 1964a

Oulimnius bertrandi Berthélemy, 1964a: 256

TYPE LOCALITY: Spain.

DISTRIBUTION: Portugal, Spain.

cyneticus Berthélemy, 1980

Oulimnius cyneticus Berthélemy, 1980: 421

Type Locality: Portugal.

Distribution: Portugal, Spain.

echinatus Berthélemy, 1979

Oulimnius echinatus Berthélemy, 1979: 77

Type Locality: Spain (Balearic Islands).

Distribution: Spain (Balearic Islands).

fuscipes (Reiche, 1879)

Limnius fuscipes Reiche, 1879: 238

Type Locality: Algeria.

Distribution: Algeria, Morocco, Spain, Tunisia.

Synonyms: *Limnius interruptus* Fairmaire, 1884: LXI[61]. — Type locality: Algeria. Note: Types not retrieved (Berthélemy 1979: 80).

Limnius kebir Alluaud, 1926: 21. — Type locality: Morocco.

hipponensis Berthélemy, 1979

Oulimnius hipponensis Berthélemy, 1979: 82

Type Locality: Algeria.

Distribution: Algeria, Tunisia.

jaechi Hernando, Ribera & Aguilera, 1998

Oulimnius jaechi Hernando, Ribera & Aguilera, 1998: 254

Type Locality: Morocco.

Distribution: Morocco.

latiusculus (Leconte, 1866)

Elmis latiuscula Leconte, 1866: 380

Type Locality: USA (Pennsylvania).

Distribution: Canada (New Brunswick, Newfoundland, Nova Scotia, Ontario, Prince Edward Island, Quebec), USA (Alabama, Connecticut, Delaware, Georgia, Indiana, Kentucky, Maryland, Mississippi, New Hampshire, New Jersey, New York, North Carolina, Pennsylvania, Rhode Island, South Carolina, Tennessee, Vermont, Virginia, West Virginia).

Note: Incorrect original spelling: "*Elmis latiusculus*"; incorrect subsequent spelling: "*Oulimnius latiuscula*" (Poole & Gentili 1996: 265); generic assignment uncertain.

major Rey, 1889

Oulimnius major Rey, 1889: 67

Type Locality: France.

Distribution: France, Netherlands, Portugal, Spain, United Kingdom.

Note: Erroneously listed as a synonym of *Oulimnius tuberculatus* (Müller, 1806a) by Mascagni & Calamandrei (1992: 131).

Synonym: *Oulimnius falcifer* Berthélemy, 1962: 218. — Type locality: France.

maurus Berthélemy, 1979

Oulimnius maurus Berthélemy, 1979: 85

Type Locality: Algeria.

Distribution: Algeria.

nitidulus (Leconte, 1866)

Elmis nitidula Leconte, 1866: 380

Type Locality: USA (New York).

Distribution: USA (Alabama, Connecticut, Delaware, Georgia, Maryland, Mississippi, New Jersey, New York, North Carolina, South Carolina, Tennessee, Virginia, West Virginia).

Note: Incorrect subsequent spellings: "*Elmis nitidulus*" (Crotch 1873: 52), "*Oulimnius nitidula*" (Poole & Gentili 1996: 265); generic assignment uncertain.

perezi (Sharp, 1872)

Limnius perezi Sharp, 1872: 264

Type Locality: Spain.

Distribution: Portugal, Spain.

reygassei (Peyerimhoff, 1929)

Limnius reygassei Peyerimhoff, 1929: 171

Type Locality: Algeria.

Distribution: Algeria (Hoggar Mountains).

Note: Not included in revisions by Berthélemy (1962, 1964a, 1979, 1980).

rivularis (Rosenhauer, 1856)

Limnius rivularis Rosenhauer, 1856: 113

Type Locality: Spain.

Distribution: Algeria, France, Germany, Italy, Morocco, Netherlands, Portugal, Spain, Tunisia, United Kingdom.

Synonyms: *Elmis subparallela* Fairmaire, 1863: 74. — Type locality: France. Note: Incorrect original spelling: "*subparallelus*".

Limnius neuter Kuwert, 1890: 20. — Type locality: Italy (Sardinia), by lectotype designation (Berthélemy 1979: 85).

troglodytes (Gyllenhal, 1827)

Limnius troglodytes Gyllenhal, 1827: 395

Type Locality: Sweden.

Distribution: Belgium, Denmark, France, Germany, Italy, Lithuania, Morocco (Chavanon et al. 2004: 158), Netherlands, Poland? (Przewoźny et al. 2011), Portugal, Spain, Sweden, Tunisia (Touaylia et al. 2010: 171), United Kingdom.

Note: Incorrect subsequent spelling: "*troglodites*" (Villa & Villa 1835: 41).

Synonyms: *Elmis fluviatilis* Stephens, 1828: 107. — Type locality: United Kingdom. Note: A male type specimen, deposited in the Natural History Museum, London, U.K., has been examined by the first author (M.A. Jäch) in 2014.

Limnius brevis Sharp, 1872: 264. — Type locality: Spain.

Limnius thermarius Sainte-Claire Deville, 1919: 263. — Type locality: France.

tuberculatus (Müller, 1806a)

Limnius tuberculatus Müller, 1806a: 199

Type Locality: Germany.

Distribution: Albania, Armenia, Austria, Belarus, Belgium, Bulgaria, Croatia (Mičetić Stanković et al. 2015: 103), Czech Republic, Denmark, Estonia, Finland, France, Germany, Georgia, Greece, Hungary, Ireland, Italy, Latvia, Lithuania, Luxembourg, Macedonia, (Montenegro), Netherlands, Norway, Poland, (Romania), Russia, Serbia, Slovakia, Slovenia, Spain, Sweden, Switzerland, Turkey, United Kingdom; specimens from Italy might belong to an undescribed species, see Čiampor & Kodada (2010).

Synonyms: *Elmis dargelasi* Latreille, 1807: 51. — Type locality: France. Note: Incorrect subsequent spelling: "*dargelasii*" (Villa & Villa 1833: 14, Dejean 1833: 131, Gemminger 1851: 10).

Elmis lacustris Stephens, 1828: 107 **syn.n.** — Type locality: United Kingdom.

Elmis variabilis Stephens, 1828: 107. — Type locality: United Kingdom.

Limnius formosus Kuwert, 1890: 44. — Type locality: France.

villosus Berthélemy, 1979

Oulimnius villosus Berthélemy, 1979: 88

Type Locality: Morocco.

Distribution: Algeria, Morocco.

Ovolara Brown, 1981c

Ovolara Brown, 1981c: 90. — Type species: *Lutrochus australis* King, 1865. — Gender feminine.

australis (King, 1865)

Lutrochus australis King, 1865: 159 ("*Lutochrus*" [Lutrochidae])

TYPE LOCALITY: Australia (New South Wales).

DISTRIBUTION: Australia (New South Wales).

leai (Carter, 1926b)

Hydrethus leai Carter, 1926b: 64

TYPE LOCALITY: Australia (Queensland).

DISTRIBUTION: Australia (Queensland).

Pachyelmis Fairmaire, 1898b

Pachyelmis Fairmaire, 1898b: 467. — Type species: *Pachyelmis validipes* Fairmaire, 1898b. — Gender feminine.

SYNONYM: *Pachelmis* Zaitzev, 1908: 309. — Type species: *Pachyelmis validipes* Fairmaire, 1898b. — Gender feminine. NOTE: Unjustified emendation of *Pachyelmis* Fairmaire, 1898b (ICZN 1999: Art. 33.2.1).

aemula Delève, 1964a

Pachyelmis aemula Delève, 1964a: 49

TYPE LOCALITY: Madagascar.

DISTRIBUTION: Madagascar.

aequata Delève, 1967c

Pachyelmis aequata Delève, 1967c: 77

TYPE LOCALITY: Congo (R).

DISTRIBUTION: Congo (R).

alluaudi Delève, 1963a

Pachyelmis alluaudi Delève, 1963a: 17

TYPE LOCALITY: Madagascar.

DISTRIBUTION: Madagascar.

amaena Grouvelle, 1906c

Pachyelmis amaena Grouvelle, 1906c: 322

TYPE LOCALITY: Tanzania.

DISTRIBUTION: Kenya, South Africa, Tanzania, Uganda? (Delève 1970a: 221: "Mt. Elgon").

NOTE: Incorrect subsequent spelling: "*amoena*" (Zaitzev 1908: 309, 1910: 37, Delève 1964c: 247, 1970a: 220).

basilewskyi Delève, 1956

Pachyelmis basilewskyi Delève, 1956: 376

Type Locality: Rwanda.

Distribution: Rwanda.

bertrandi Delève, 1964a

Pachyelmis bertrandi Delève, 1964a: 52

Type Locality: Madagascar.

Distribution: Madagascar.

bigibbulosa Delève, 1973b

Pachyelmis bigibbulosa Delève, 1973b: 310

Type Locality: Liberia.

Distribution: Liberia.

capuroni Delève, 1963a

Pachyelmis capuroni Delève, 1963a: 12

Type Locality: Madagascar.

Distribution: Madagascar.

collarti Delève, 1964c

Pachyelmis collarti Delève, 1964c: 245

Type Locality: Congo (DR).

Distribution: Congo (DR), Congo (R).

convexa convexa Grouvelle, 1911c

Pachyelmis convexa Grouvelle, 1911c: 274

Type Locality: Uganda.

Distribution: Congo (DR), Kenya, South Africa, Uganda, Zimbabwe.

convexa janssensi Delève, 1964c

Pachyelmis convexa janssensi Delève, 1964c: 243

Type Locality: Tanzania.

Distribution: Tanzania.

distinguenda Delève, 1938

Pachyelmis distinguenda Delève, 1938: 371

Type Locality: Congo (DR).

Distribution: Congo (DR).

fairmairei (Grouvelle, 1899)

Helmis fairmairei Grouvelle, 1899: 184

TYPE LOCALITY: Madagascar.

DISTRIBUTION: Madagascar.

gibba Grouvelle, 1911c

Pachyelmis gibba Grouvelle, 1911c: 275

TYPE LOCALITY: Uganda.

DISTRIBUTION: Congo (DR), Uganda.

grouvellei (Zaitzev, 1908)

Helmis grouvellei Zaitzev, 1908: 302

TYPE LOCALITY: Madagascar.

DISTRIBUTION: Madagascar.

NOTE: Substitute name for *Elmis subsulcata* Fairmaire, 1897b.

HOMONYM: *Elmis subsulcata* Fairmaire, 1897b: 369. NOTE: Incorrect original spelling: "*subsulcatus*"; formerly a junior secondary homonym of *Helmis subsulcata* Grouvelle, 1889b: 403 (*Hexacylloepus*); permanently invalid (ICZN 1999: Art. 59.3).

ingens Grouvelle, 1906a

Pachyelmis ingens Grouvelle, 1906a: 158

TYPE LOCALITY: Madagascar.

DISTRIBUTION: Madagascar.

interstitialis interstitialis Fairmaire, 1902

Pachyelmis interstitialis Fairmaire, 1902: 344

TYPE LOCALITY: Madagascar.

DISTRIBUTION: Madagascar.

interstitialis meridionalis Delève, 1963a

Pachyelmis interstitialis meridionalis Delève, 1963a: 5

TYPE LOCALITY: Madagascar.

DISTRIBUTION: Madagascar.

madudana Delève, 1937a

Pachyelmis madudana Delève, 1937a: 159

TYPE LOCALITY: Congo (DR).

DISTRIBUTION: Cameroon, Congo (DR), Congo (R), Ivory Coast.

manca Delève, 1938

Pachyelmis manca Delève, 1938: 371

TYPE LOCALITY: Congo (DR).

DISTRIBUTION: Congo (DR).

obesa Delève, 1964a

Pachyelmis obesa Delève, 1964a: 51

Type Locality: Madagascar.

Distribution: Madagascar.

obliqua Grouvelle, 1906a

Pachyelmis obliqua Grouvelle, 1906a: 159

Type Locality: Madagascar.

Distribution: Madagascar.

obscura minor Delève, 1963a

Pachyelmis obscura minor Delève, 1963a: 16

Type Locality: Madagascar.

Distribution: Madagascar.

obscura obscura Delève, 1963a

Pachyelmis obscura Delève, 1963a: 15

Type Locality: Madagascar.

Distribution: Madagascar.

persimilis Delève, 1974

Pachyelmis persimilis Delève, 1974: 284

Type Locality: Ghana.

Distribution: Ghana.

quadricarinata Delève, 1964a

Pachyelmis quadricarinata Delève, 1964a: 48

Type Locality: Madagascar.

Distribution: Madagascar.

regimbarti Grouvelle, 1906a

Pachyelmis regimbarti Grouvelle, 1906a: 158

Type Locality: Madagascar.

Distribution: Madagascar.

rubripes (Fairmaire, 1898a)

Elmis rubripes Fairmaire, 1898a: 225

Type Locality: Madagascar.

Distribution: Madagascar.

rufomarginata Delève, 1964c

Pachyelmis rufomarginata Delève, 1964c: 251

Type Locality: Congo (DR).

Distribution: Angola, Congo (DR).

rufula Delève, 1963a

Pachyelmis rufula Delève, 1963a: 8

Type Locality: Madagascar.

Distribution: Madagascar.

schoutedeni Delève, 1938

Pachyelmis schoutedeni Delève, 1938: 370

Type Locality: Congo (DR).

Distribution: Congo (DR).

securigera Delève, 1974

Pachyelmis securigera Delève, 1974: 285

Type Locality: Ghana.

Distribution: Ghana.

silvatica Grouvelle, 1906a

Pachyelmis silvatica Grouvelle, 1906a: 157

Type Locality: Madagascar.

Distribution: Madagascar.

tibialis Delève, 1968b

Pachyelmis tibialis Delève, 1968b: 206

Type Locality: Ivory Coast.

Distribution: Ivory Coast.

upembana Delève, 1955

Pachyelmis upembana Delève, 1955: 20

Type Locality: Congo (DR).

Distribution: Congo (DR).

validipes Fairmaire, 1898b

Pachyelmis validipes Fairmaire, 1898b: 468

Type Locality: Madagascar.

Distribution: Madagascar.

Pagelmis Spangler, 1981b

Pagelmis Spangler, 1981b: 286. — Type species: *Pagelmis amazonica* Spangler, 1981b. — Gender feminine.

amazonica Spangler, 1981b

Pagelmis amazonica Spangler, 1981b: 287

Type Locality: Ecuador.

Distribution: Ecuador, Suriname.

Paramacronychus Nomura, 1958b

Paramacronychus Nomura, 1958b: 47. — Type species: *Paramacronychus granulatus* Nomura, 1958b. — Gender masculine.

crassipes (Champion, 1927)

Zaitzevia crassipes Champion, 1927: 50

Type Locality: India (Himachal Pradesh).

Distribution: India (Himachal Pradesh).

Note: Transferred to *Paramacronychus* by Satô (1976).

granulatus Nomura, 1958b

Paramacronychus granulatus Nomura, 1958b: 48

Type Locality: Japan (Honshu).

Distribution: Japan (Honshu, Shikoku, Yakushima).

Note: Incorrect subsequent spelling: "*granulates*" (Ogawa 2013: 65).

Parapotamophilus Brown, 1981c

Parapotamophilus Brown, 1981c: 94. — Type species: *Parapotamophilus gressitti* Brown, 1981c. — Gender masculine.

gressitti Brown, 1981c

Parapotamophilus gressitti Brown, 1981c: 96

Type Locality: Papua New Guinea.

Distribution: Papua New Guinea.

Peloriolus Delève, 1964b

Peloriolus Delève, 1964b: 161. — Type species: *Peloriolus granulosus* Delève, 1964b. — Gender masculine.

brunneus (F.H. Waterhouse, 1879)

Elmis brunnea F.H. Waterhouse, 1879: 532 **comb.n.**

Type Locality: Saint Helena (British Overseas Territory).

Distribution: Saint Helena (British Overseas Territory).

Note: Incorrect original spelling: "*Elmis brunneus*"; transferred herewith from *Elmis* Latreille, 1802.

costulatipennis Delève, 1964b

Peloriolus costulatipennis Delève, 1964b: 165

Type Locality: South Africa.

Distribution: South Africa.

difficilis Delève, 1970a

Peloriolus difficilis Delève, 1970a: 223

Type Locality: South Africa.

Distribution: South Africa.

granulosus Delève, 1964b

Peloriolus granulosus Delève, 1964b: 162

Type Locality: South Africa.

Distribution: Angola, South Africa.

interstitialis Delève, 1970a

Peloriolus interstitialis Delève, 1970a: 225

Type Locality: South Africa.

Distribution: South Africa.

parvulus Delève, 1964b

Peloriolus parvulus Delève, 1964b: 164

Type Locality: South Africa.

Distribution: South Africa.

patruelis Delève, 1966b

Peloriolus patruelis Delève, 1966b: 105

Type Locality: South Africa.

Distribution: Angola, South Africa.

pilosellus Delève, 1966b

Peloriolus pilosellus Delève, 1966b: 107

TYPE LOCALITY: South Africa.

DISTRIBUTION: South Africa.

Phanoceroides Hinton, 1939a

Phanoceroides Hinton, 1939a: 169. — Type species: *Phanoceroides aquaticus* Hinton, 1939a. — Gender masculine.

aquaticus Hinton, 1939a

Phanoceroides aquaticus Hinton, 1939a: 172

TYPE LOCALITY: Brazil (Amazonas); erroneously, Hinton (1939a: 173) attributed the type locality to the State of Pará.

DISTRIBUTION: Brazil (Amazonas).

NOTE: Incorrect subsequent spelling: "*acautica*" (Manzo 2005: 203).

Phanocerus Sharp, 1882

Phanocerus Sharp, 1882: 128. — Type species: *Phanocerus clavicornis* Sharp, 1882. — Gender masculine.

bugnioni Grouvelle, 1902b

Phanocerus bugnioni Grouvelle, 1902b: 466

TYPE LOCALITY: Colombia.

DISTRIBUTION: Colombia.

charopus Spangler, 1966

Phanocerus charopus Spangler, 1966: 413

TYPE LOCALITY: Peru.

DISTRIBUTION: Peru.

clavicornis Sharp, 1882

Phanocerus clavicornis Sharp, 1882: 129

TYPE LOCALITY: Guatemala.

DISTRIBUTION: Belize, Brazil (Amazonas, Goiás, Rio de Janeiro, Santa Catarina), Costa Rica, Cuba, Dominican Republic, Guatemala, Haiti, Honduras, Jamaica, Mexico, Nicaragua, Panama, Puerto Rico, USA (Texas), Venezuela.

SYNONYMS: *Phanocerus hubbardi* Schaeffer, 1911: 119. — Type locality: Jamaica.

Phanocerus helmoides Darlington, 1936: 74. — Type locality: Haiti.

congener Grouvelle, 1898a

Phanocerus congener Grouvelle, 1898a: 46

Type Locality: Grenada.

Distribution: Belize, Costa Rica, Ecuador, Grenada, Guatemala, Nicaragua, Panama, Trinidad and Tobago (Tobago, Trinidad), Venezuela.

Note: Incorrect subsequent spelling: "*conger*" (Shepard 2004: 55).

rufus Maier, 2013

Phanocerus rufus Maier, 2013: 72

Type Locality: Venezuela.

Distribution: Venezuela.

sharpi Grouvelle, 1896b

Phanocerus sharpi Grouvelle, 1896b: 4 [(8)]

Type Locality: Uruguay.

Distribution: Argentina, Paraguay, Uruguay.

Pharceonus Spangler & Santiago-Fragoso, 1992

Pharceonus Spangler & Santiago-Fragoso, 1992: 23. — Type species: *Pharceonus volcanus* Spangler & Santiago-Fragoso, 1992. — Gender masculine.

ariasi Maier, 2013

Pharceonus ariasi Maier, 2013: 75

Type Locality: Venezuela.

Distribution: Venezuela.

cianferonii Monte & Mascagni, 2012

Pharceonus cianferonii Monte & Mascagni, 2012: 26

Type Locality: Ecuador.

Distribution: Ecuador.

grandis Maier, 2013

Pharceonus grandis Maier, 2013: 78

Type Locality: Venezuela.

Distribution: Venezuela.

volcanus Spangler & Santiago-Fragoso, 1992

Pharceonus volcanus Spangler & Santiago-Fragoso, 1992: 23

Type Locality: Panama.

Distribution: Colombia, Costa Rica, Ecuador, Panama, Venezuela.

Pilielmis Hinton, 1971b

Pilielmis Hinton, 1971b: 161. — Type species: *Pilielmis halia* Hinton, 1971b. — Gender feminine.

abdera Hinton, 1971b

Pilielmis abdera Hinton, 1971b: 166

TYPE LOCALITY: Brazil (Rondônia); erroneously, Hinton (1971b: 166) attributed the type locality to Mato Grosso ("Mato Grosso: Porto Velho, vii–ix. 1937 (H.E. Hinton)").

DISTRIBUTION: Brazil (Rondônia); erroneously recorded from Mato Grosso by Segura et al. (2013: 38) and erroneously recorded from Amazonas (Manaus) and Pará by Passos et al. (2010a: 541); this species is known only from the type locality.

apama Hinton, 1971b

Pilielmis apama Hinton, 1971b: 165

TYPE LOCALITY: French Guiana.

DISTRIBUTION: Colombia, French Guiana.

clita Hinton, 1971b

Pilielmis clita Hinton, 1971b: 165

TYPE LOCALITY: Brazil (Amazonas); erroneously, Hinton (1971b: 166) attributed the type locality (Manaus) to the State of Pará.

DISTRIBUTION: Brazil (Amazonas); erroneously recorded from Pará by Segura et al. (2013: 38) and Passos et al. (2010a: 541); one of the 63 paratypes, housed in the Natural History Museum, London, U.K., is labelled "Belem" (State of Pará), however, according to M. Barclay (email of 27.XII.2013), this is a "female, missing head, thorax and one elytron. It is probably a specimen of *P. halia* that was incorrectly sorted and labelled by Hinton".

halia Hinton, 1971b

Pilielmis halia Hinton, 1971b: 164

TYPE LOCALITY: Brazil (Pará).

DISTRIBUTION: Brazil (Pará).

murcia Hinton, 1971b

Pilielmis murcia Hinton, 1971b: 163

TYPE LOCALITY: French Guiana.

DISTRIBUTION: Brazil (Amazonas, Pará), French Guiana.

sara Hinton, 1971b

Pilielmis sara Hinton, 1971b: 164

TYPE LOCALITY: Brazil (Amazonas).

DISTRIBUTION: Brazil (Amazonas).

Podelmis Hinton, 1941b

Podelmis Hinton, 1941b: 68. — Type species: *Elmis coronata* Champion, 1923. — Gender feminine.

aenea Delève, 1973a

Podelmis aenea Delève, 1973a: 20

Type Locality: Sri Lanka.

Distribution: Sri Lanka.

atra Jäch, 1982b

Podelmis atra Jäch, 1982b: 98

Type Locality: Sri Lanka.

Distribution: Sri Lanka.

Note: Incorrect original spelling: "*ater*", herewith changed mandatorily (ICZN 1999: Art. 34.2).

coronata (Champion, 1923)

Elmis coronata Champion, 1923: 172

Type Locality: India (Tamil Nadu).

Distribution: India (Tamil Nadu).

cruzei Jäch, 1982b

Podelmis cruzei Jäch, 1982b: 99

Type Locality: Sri Lanka.

Distribution: Sri Lanka.

graphica Jäch, 1982b

Podelmis graphica Jäch, 1982b: 100

Type Locality: Sri Lanka.

Distribution: Sri Lanka.

humeralis Jäch, 1982b

Podelmis humeralis Jäch, 1982b: 100

Type Locality: Sri Lanka.

Distribution: Sri Lanka.

metallica Delève, 1973a

Podelmis metallica Delève, 1973a: 22

Type Locality: Sri Lanka.

Distribution: Sri Lanka.

ovalis Jäch, 1982b

Podelmis ovalis Jäch, 1982b: 101

Type Locality: Sri Lanka.

Distribution: Sri Lanka.

palniensis (Champion, 1923)

Elmis palniensis Champion, 1923: 173

Type Locality: India (Tamil Nadu).

Distribution: India (Tamil Nadu).

quadriplagiata (Motschulsky, 1860)

Ancyronyx quadriplagiatus Motschulsky, 1860: 47

Type Locality: Sri Lanka.

Distribution: Sri Lanka.

Synonym: *Leptelmis nietneri* Champion, 1923: 168. — Type locality: Sri Lanka.

similis Jäch, 1982b

Podelmis similis Jäch, 1982b: 102

Type Locality: Sri Lanka.

Distribution: Sri Lanka.

torrentium (Champion, 1923)

Elmis torrentium Champion, 1923: 173

Type Locality: India (Tamil Nadu).

Distribution: India (Tamil Nadu).

Note: Epithet name is a Latin participle (genitive plural of torrens [stream]), identical in all three genders.

v-altum (Champion, 1923)

Elmis v-altum Champion, 1923: 174

Type Locality: India (Tamil Nadu).

Distribution: India (Tamil Nadu).

Note: Epithet name is a noun in apposition.

viridiaenea Jäch, 1982b

Podelmis viridiaenea Jäch, 1982b: 102

Type Locality: Sri Lanka.

Distribution: Sri Lanka.

xanthogramma Jäch, 1983a

Podelmis xanthogramma Jäch, 1983a: 111

Type Locality: Malaysia (Selangor).

Distribution: Malaysia (Selangor).

Note: Combines characters of *Ancyronyx* and *Podelmis*.

Podonychus Jäch & Kodada, 1997

Podonychus Jäch & Kodada, 1997: 18. — Type species: *Podonychus sagittarius* Jäch & Kodada, 1997. — Gender masculine.

sagittarius Jäch & Kodada, 1997

Podonychus sagittarius Jäch & Kodada, 1997: 23

Type Locality: Indonesia (Siberut).

Distribution: Indonesia (Siberut).

Portelmis Sanderson, 1953b

Portelmis Sanderson, 1953b: 35. — Type species: *Stenelmis nevermanni* Hinton, 1936c. — Gender feminine.

guianensis Przewoźny & Fernandes, 2012

Portelmis guianensis Przewoźny & Fernandes, 2012: 59

Type Locality: French Guiana.

Distribution: French Guiana.

gurneyi Spangler, 1980a

Portelmis gurneyi Spangler, 1980a: 63

Type Locality: Ecuador.

Distribution: Brazil (Amazonas), Colombia, Ecuador, Peru.

kinonatilis Fernandes, Passos & Hamada, 2010b

Portelmis kinonatilis Fernandes, Passos & Hamada, 2010b: 34

Type Locality: Brazil (Amazonas).

Distribution: Brazil (Amazonas).

nevermanni (Hinton, 1936c)

Stenelmis nevermanni Hinton, 1936c: 424

Type Locality: Costa Rica.

Distribution: Costa Rica.

paulicruzi Fernandes, Passos & Hamada, 2010b

Portelmis paulicruzi Fernandes, Passos & Hamada, 2010b: 40

Type Locality: Brazil (Amazonas).

Distribution: Brazil (Amazonas).

Potamocares Grouvelle, 1920

Potamocares Grouvelle, 1920: 198. — Type species: *Potamocares striatus* Grouvelle, 1920. — Gender masculine.

burgeoni (Delève, 1937b)

Hydrethus burgeoni Delève, 1937b: 88

Type Locality: Congo (DR).

Distribution: Congo (DR).

jeanneli (Hinton, 1937c)

Hydrethus jeanneli Hinton, 1937c: 302

Type Locality: Mozambique.

Distribution: Mozambique.

marlieri Delève, 1963c

Potamocares marlieri Delève, 1963c: 446

Type Locality: Congo (DR).

Distribution: Congo (DR).

striatus Grouvelle, 1920

Potamocares striatus Grouvelle, 1920: 199

Type Locality: Kenya.

Distribution: Kenya.

Potamodytes Grouvelle, 1896c

Potamodytes Grouvelle, 1896c: 78. — Type species: *Potamophilus abdominalis* C.O. Waterhouse, 1879. — Gender masculine.

Note: Incorrect subsequent spelling: "*Potamodites*" (Grouvelle 1913: 569).

Synonym: *Gridelliana* Bollow, 1939: 149. — Type species: *Gridelliana zavattarii* Bollow, 1939. — Gender feminine.

abdominalis (C.O. Waterhouse, 1879)

Potamophilus abdominalis C.O. Waterhouse, 1879: 529

Type Locality: Madagascar.

Distribution: Madagascar.

africanus (Boheman, 1851)

Potamophilus africanus Boheman, 1851: 585

TYPE LOCALITY: South Africa.

DISTRIBUTION: Widely distributed in Africa, from Sudan to South Africa (Delève 1967b: 429, Więźlak 1987b: 449), including Comoros and Mayotte (French Overseas Department).

SYNONYM: *Potamodytes schoutedeni* Delève, 1937b: 98. — Type locality: Congo (DR).

alluaudi Grouvelle, 1920

Potamodytes alluaudi Grouvelle, 1920: 196

TYPE LOCALITY: Kenya.

DISTRIBUTION: Cameroon, Congo (DR), Ethiopia, Ghana, Kenya, Rwanda, Uganda.

SYNONYM: *Potamodytes ambiguus* Delève, 1937b: 94. — Type locality: Congo (DR).

angustatus Hinton, 1937c

Potamodytes angustatus Hinton, 1937c: 297

TYPE LOCALITY: Uganda.

DISTRIBUTION: South Africa, Uganda, Zambia.

antennatus (Dohrn, 1882)

Potamophilus antennatus Dohrn, 1882: 251

TYPE LOCALITY: Ghana; the type locality, "Akem, Guinea" (Dohrn 1882: 251), obviously refers to the former "District of Akem" in Ghana (see Hay 1876).

DISTRIBUTION: Ghana, Guinea?, Ivory Coast, Liberia; literature records from Guinea obviously refer to the original description, however, terms like "Guineaküste" (Guinea Coast) (p. 250), and "Guinea" (p. 251), during these days comprised the entire coast of West Africa and did not specifically refer to the present state of Guinea.

apicalis Hinton, 1937c

Potamodytes apicalis Hinton, 1937c: 299

TYPE LOCALITY: Nigeria.

DISTRIBUTION: Nigeria.

bispinosus Grouvelle, 1914

Potamodytes bispinosus Grouvelle, 1914: 201

TYPE LOCALITY: Tanzania.

DISTRIBUTION: Cameroon, Tanzania.

bisulcatus Delève, 1962

Potamodytes bisulcatus Delève, 1962: 10

TYPE LOCALITY: Madagascar.

DISTRIBUTION: Madagascar.

brincki Delève, 1970a

Potamodytes brincki Delève, 1970a: 219

TYPE LOCALITY: South Africa.

DISTRIBUTION: South Africa.

cernuus Delève, 1962

Potamodytes cernuus Delève, 1962: 9

TYPE LOCALITY: Madagascar.

DISTRIBUTION: Madagascar.

convexior Delève, 1937b

Potamodytes convexior Delève, 1937b: 97

TYPE LOCALITY: Congo (DR).

DISTRIBUTION: Angola, Congo (DR).

cribricollis Delève, 1937b

Potamodytes cribricollis Delève, 1937b: 102

TYPE LOCALITY: Congo (DR).

DISTRIBUTION: Congo (DR).

delusor Delève, 1962

Potamodytes delusor Delève, 1962: 7

TYPE LOCALITY: Madagascar.

DISTRIBUTION: Madagascar.

descarpentriesi Delève, 1967a

Potamodytes descarpentriesi Delève, 1967a: 321

TYPE LOCALITY: Congo (R).

DISTRIBUTION: Congo (R), Gabon.

ghesquierei Delève, 1937b

Potamodytes ghesquierei Delève, 1937b: 104

TYPE LOCALITY: Congo (DR).

DISTRIBUTION: Congo (DR).

NOTE: Incorrect original spelling: "*ghesquièrei*".

grouvellei Delève, 1937b

Potamodytes grouvellei Delève, 1937b: 105

TYPE LOCALITY: Congo (DR).

DISTRIBUTION: Congo (DR), Congo (R), Gabon.

hastatus Delève, 1945c

Potamodytes hastatus Delève, 1945c: 152

Type Locality: Guinea.

Distribution: Cameroon, Congo (R), Guinea, Ivory Coast.

janssensi Delève, 1955

Potamodytes janssensi Delève, 1955: 18

Type Locality: Congo (DR).

Distribution: Congo (DR).

latus Grouvelle, 1906a

Potamodytes latus Grouvelle, 1906a: 145

Type Locality: Madagascar.

Distribution: Madagascar.

lokis Więźlak, 1987c

Potamodytes lokis Więźlak, 1987c: 119

Type Locality: Tanzania.

Distribution: Tanzania.

major Kolbe, 1898

Potamodytes major Kolbe, 1898: 127

Type Locality: Congo (DR).

Distribution: Congo (DR), Congo (R), Rwanda, Uganda, Zambia.

Synonym: *Potamodytes major invalidus* Delève, 1937b: 97. — Type locality: Congo (DR). Note: Originally described as "*Potamodytes major invalidus* var. nov."; might be a good species.

marshalli Hinton, 1937c

Potamodytes marshalli Hinton, 1937c: 292

Type Locality: Sierra Leone.

Distribution: Ghana, Ivory Coast, Sierra Leone.

mucronatus Delève, 1937a

Potamodytes mucronatus Delève, 1937a: 149

Type Locality: Congo (DR).

Distribution: Congo (DR), Congo (R), Gabon.

overlaeti Delève, 1937b

Potamodytes overlaeti Delève, 1937b: 92

TYPE LOCALITY: Congo (DR).

DISTRIBUTION: Angola, Cameroon, Congo (DR), Ghana.

oxypterus (Fairmaire, 1889b)

Potamophilus oxypterus Fairmaire, 1889b: XC[90]

TYPE LOCALITY: Madagascar.

DISTRIBUTION: Madagascar.

NOTE: Redescribed by Delève (1962: 2).

perrieri Grouvelle, 1906a

Potamodytes perrieri Grouvelle, 1906a: 145

TYPE LOCALITY: Madagascar.

DISTRIBUTION: Madagascar.

praeteritus Delève, 1967b

Potamodytes praeteritus Delève, 1967b: 431

TYPE LOCALITY: Madagascar.

DISTRIBUTION: Madagascar.

puncticollis Delève, 1966d

Potamodytes puncticollis Delève, 1966d: 46

TYPE LOCALITY: Angola.

DISTRIBUTION: Angola.

sericeus sericeus Delève, 1945c

Potamodytes sericeus Delève, 1945c: 154

TYPE LOCALITY: Congo (DR).

DISTRIBUTION: Congo (DR).

sericeus variipennis Delève, 1945c

Potamodytes sericeus variipennis Delève, 1945c: 155

TYPE LOCALITY: Congo (DR).

DISTRIBUTION: Congo (DR).

spangleri Delève, 1973b

Potamodytes spangleri Delève, 1973b: 300

TYPE LOCALITY: Liberia.

DISTRIBUTION: Liberia.

spinosus Grouvelle, 1913

Potamodytes spinosus Grouvelle, 1913: 572

TYPE LOCALITY: Congo (R).

DISTRIBUTION: Angola? (Delève 1966d: 46), Cameroon, Congo (DR), Congo (R), Tanzania.

SYNONYM: *Potamodytes bidentatus* Delève, 1937b: 99. — Type locality: Congo (DR).

subrotundatus Pic, 1939

Potamodytes subrotundatus Pic, 1939: 143

TYPE LOCALITY: Egypt.

DISTRIBUTION: Egypt, Israel, Saudi Arabia, United Arab Emirates, Yemen.

SYNONYMS: *Potamodytes ochus* Hinton, 1948: 137. — Type locality: Yemen.

Potamodytes pici Delève, 1967b: 431. NOTE: Substitute name for *Potamodytes bispinosus* Pic, 1950.

HOMONYM: *Potamodytes bispinosus* Pic, 1950: 23. — Type locality: Egypt. NOTE: Junior primary homonym of *Potamodytes bispinosus* Grouvelle, 1914.

tarnawskii Więźlak, 1987b

Potamodytes tarnawskii Więźlak, 1987b: 447

TYPE LOCALITY: Cameroon.

DISTRIBUTION: Cameroon.

tuberosus Hinton, 1937c

Potamodytes tuberosus Hinton, 1937c: 294

TYPE LOCALITY: Gabon.

DISTRIBUTION: Cameroon, Central African Republic, Congo (R), Gabon, Liberia, Nigeria.

zavattarii (Bollow, 1939)

Gridelliana zavattarii Bollow, 1939: 150

TYPE LOCALITY: Ethiopia.

DISTRIBUTION: Ethiopia, Somalia.

Potamogethes Delève, 1963c

Potamogethes Delève, 1963c: 439. — Type species: *Hydrethus major* Delève, 1937b. — Gender masculine.

NOTE: The masculine gender is fixed herewith; the second component of the genus name, -gethes, is derived from the Greek word Gethos (το γῆθος, English: joy), which is neuter; however, since the ending is not the original one (ICZN 1999: Art. 30.1.2) and not a Latin gender ending (ICZN 1999: Art. 30.1.3), and since the gender was not fixed in the original description, we decided to treat it as masculine in order to agree with numerous other Coleoptera genera based on the Greek word Gethos, e.g. *Meligethes* Stephens, 1830.

crassipes Delève, 1963c

Potamogethes crassipes Delève, 1963c: 441

Type Locality: Angola.

Distribution: Angola, Central African Republic, Congo (R).

cribricollis Delève, 1963c

Potamogethes cribricollis Delève, 1963c: 443

Type Locality: Congo (DR).

Distribution: Angola, Congo (DR).

danielssoni Więźlak, 1987b

Potamogethes danielssoni Więźlak, 1987b: 442

Type Locality: Cameroon.

Distribution: Cameroon.

ivorensis Delève, 1968b

Potamogethes ivorensis Delève, 1968b: 197

Type Locality: Ivory Coast.

Distribution: Gabon, Ghana, Ivory Coast.

major (Delève, 1937b)

Hydrethus major Delève, 1937b: 90

Type Locality: South Sudan.

Distribution: Congo (DR), South Sudan.

Potamolatres Delève, 1963c

Potamolatres Delève, 1963c: 454. — Type species: *Potamolatres costulatus* Delève, 1963c. — Gender masculine.

costulatus Delève, 1963c

Potamolatres costulatus Delève, 1963c: 454

Type Locality: Madagascar.

Distribution: Madagascar.

Potamophilinus Grouvelle, 1896c

Potamophilinus Grouvelle, 1896c: 78. — Type species: *Potamophilus longipes* Grouvelle, 1892b. — Gender masculine.

SYNONYM: *Freyiella* Bollow, 1938: 168. — Type species: *Freyiella foveicollis* Bollow, 1938. — Gender feminine.

bispinosus (Bollow, 1938)

Freyiella bispinosa Bollow, 1938: 172

TYPE LOCALITY: China (Yunnan).

DISTRIBUTION: China (Yunnan).

costatus Hinton, 1935

Potamophilinus costatus Hinton, 1935: 174

TYPE LOCALITY: Sri Lanka.

DISTRIBUTION: Sri Lanka.

NOTE: Incorrect subsequent spelling: "*costataus*" (Satô 2002: 166).

foveicollis (Bollow, 1938)

Freyiella foveicollis Bollow, 1938: 169

TYPE LOCALITY: China (Guangdong).

DISTRIBUTION: China (Guangdong), Vietnam.

gravastellus Hinton, 1936e

Potamophilinus gravastellus Hinton, 1936e: 216

TYPE LOCALITY: Malaysia (Sarawak).

DISTRIBUTION: Brunei, Malaysia (Sarawak).

impressicollis Delève, 1973a

Potamophilinus impressicollis Delève, 1973a: 7

TYPE LOCALITY: Sri Lanka.

DISTRIBUTION: Sri Lanka.

javanicus (Coquerel, 1851)

Potamophilus javanicus Coquerel, 1851: 598

TYPE LOCALITY: Indonesia (Java).

DISTRIBUTION: Indonesia (Java, Sumatra).

longipes (Grouvelle, 1892b)

Potamophilus longipes Grouvelle, 1892b: 864

TYPE LOCALITY: Myanmar.

DISTRIBUTION: Myanmar.

orientalis (Guérin-Méneville, 1835)

Potamophilus orientalis Guérin-Méneville, 1835: pl. 20

Type Locality: Indonesia (Java).

Distribution: Indonesia (Bali, Java, Lombok).

Note: Authorship ascribed to Gory in original description; name available according to ICZN (1999: Art. 12.2.7), "indication" provided by fig. 1 in the original description; formal description published by Gory (1844: 70).

Homonym: *Potamophilus orientalis* Castelnau, 1840: 41. — Type locality: Indonesia (Java). Note: Primary homonym (and probably a junior synonym) of *Potamophilinus orientalis* (Guérin-Méneville, 1835); permanently invalid.

papuanus irianus Satô, 1973

Potamophilinus papuanus irianus Satô, 1973: 470

Type Locality: Indonesia (Papua).

Distribution: Indonesia (Papua).

papuanus papuanus Satô, 1973

Potamophilinus papuanus Satô, 1973: 469

Type Locality: Papua New Guinea.

Distribution: Papua New Guinea.

perplexus (Waterhouse, 1876)

Potamophilus perplexus Waterhouse, 1876: 18

Type Locality: Indonesia (Java).

Distribution: Indonesia (Java).

sumatrensis Delève, 1967b

Potamophilinus sumatrensis Delève, 1967b: 424

Type Locality: Indonesia (Sumatra).

Distribution: Indonesia (Sumatra); records from Papua New Guinea (Bismarck Archipelago) by Delève (1973c: 28) are obviously based on misidentification or incorrect labelling.

torrenticola Jäch, 1982b

Potamophilinus torrenticola Jäch, 1982b: 94

Type Locality: Sri Lanka.

Distribution: Sri Lanka.

tuberculatus Hinton, 1935

Potamophilinus tuberculatus Hinton, 1935: 175

Type Locality: Sri Lanka.

Distribution: Sri Lanka.

Potamophilops Grouvelle, 1896c

Potamophilops Grouvelle, 1896c: 78. — Type species: *Potamophilus cinereus* Blanchard, 1841. — Gender masculine.

bostrychophallus Maier, 2013

Potamophilops bostrychophallus Maier, 2013: 83

TYPE LOCALITY: Venezuela.

DISTRIBUTION: Venezuela.

bragaorum Fernandes & Hamada, 2012

Potamophilops bragaorum Fernandes & Hamada, 2012: 719

TYPE LOCALITY: Brazil (Tocantins).

DISTRIBUTION: Brazil (Tocantins).

cinereus (Blanchard, 1841)

Potamophilus cinereus Blanchard, 1841: 60

TYPE LOCALITY: Argentina.

DISTRIBUTION: Argentina, Bolivia, Brazil (Mato Grosso, São Paulo), Paraguay.

Potamophilus Germar, 1811

Potamophilus Germar, 1811: 41. — Type species: *Parnus acuminatus* Fabricius, 1792. — Gender masculine.

NOTE: Incorrect subsequent spelling: "*Potamophylus*" (Dufour 1834: 64).

SYNONYM: *Hydera* Latreille, 1816: 268. — Type species: *Parnus acuminatus* Fabricius, 1792. — Gender feminine.

acuminatus (Fabricius, 1792)

Parnus acuminatus Fabricius, 1792: 246

TYPE LOCALITY: Germany.

DISTRIBUTION: Afghanistan, Albania (M. Hess, pers. comm.), Austria, Azerbaidzhan, Belarus, Bulgaria, Croatia, Czech Republic, France, Germany, Greece, Hungary, Israel, Italy, Lebanon, Netherlands, Poland, Romania, Russia, Serbia, Slovakia, Slovenia, Spain, Syria, Tunisia (Touaylia et al. 2010: 172), Turkey, Turkmenistan, Ukraine.

albertisii Grouvelle, 1896d

Potamophilus albertisii Grouvelle, 1896d: 32

TYPE LOCALITY: Papua New Guinea.

DISTRIBUTION: Indonesia (Papua), Papua New Guinea.

NOTE: Incorrect subsequent spelling: "*albertisi*" (Zaitzev 1908: 287, 1910: 6).

feae Grouvelle, 1892b

Potamophilus feae Grouvelle, 1892b: 864

Type Locality: Myanmar.

Distribution: Myanmar.

Note: Incorrect subsequent spelling: "*feai*" (Zaitzev 1908: 287, 1910: 6).

papuanus (Carter, 1930)

Stetholus papuanus Carter, 1930: 189

Type Locality: Papua New Guinea.

Distribution: Papua New Guinea.

spinicollis Delève, 1968c

Potamophilus spinicollis Delève, 1968c: 151

Type Locality: Vietnam.

Distribution: Vietnam.

Prionosolus Jäch & Kodada, 1997

Prionosolus Jäch & Kodada, 1997: 9. — Type species: *Prionosolus venatorcapitis* Jäch & Kodada, 1997. — Gender masculine.

bobae Jäch & Kodada, 1997

Prionosolus bobae Jäch & Kodada, 1997: 18

Type Locality: Philippines (Luzon).

Distribution: Philippines (Luzon).

ciampori Jäch & Kodada, 1997

Prionosolus ciampori Jäch & Kodada, 1997: 17

Type Locality: Malaysia (Sabah).

Distribution: Malaysia (Sabah).

palawanus Freitag, 2008

Prionosolus palawanus Freitag, 2008: 298

Type Locality: Philippines (Palawan).

Distribution: Philippines (Palawan).

venatorcapitis Jäch & Kodada, 1997

Prionosolus venatorcapitis Jäch & Kodada, 1997: 11

Type Locality: Malaysia (Sarawak).

Distribution: Indonesia (Siberut)?, Malaysia (Sabah?, Sarawak).

Promoresia Sanderson, 1953a

Promoresia Sanderson, 1953a: 155. — Type species: *Helmis tardella* Fall, 1925. — Gender feminine.

NOTE: Originally described in a key; formal description published by Sanderson (1954: 9); very similar to *Heterlimnius* Hinton, 1935 and *Optioservus* Sanderson, 1953a.

elegans (Le Conte, 1852)

Limnius elegans Le Conte, 1852: 43

TYPE LOCALITY: USA (Massachusetts).

DISTRIBUTION: Canada (Quebec), USA (Alabama, Connecticut, Kentucky, Maryland, Massachusetts, New Jersey, New York, North Carolina, Pennsylvania, Tennessee, Virginia, West Virginia).

SYNONYM: *Limnius subarcticus* Brown, 1930a: 241. — Type locality: Canada (Quebec).

tardella (Fall, 1925)

Helmis tardella Fall, 1925: 179

TYPE LOCALITY: USA (Massachusetts).

DISTRIBUTION: Canada (Labrador, New Brunswick, Newfoundland, Nova Scotia, Quebec), USA (Maine, Maryland, Massachusetts, New Hampshire, New Jersey, New York, Pennsylvania, Virginia, West Virginia).

NOTE: Incorrect original spelling: "*Helmis tardellus*".

Pseudamophilus Bollow, 1940b

Pseudamophilus Bollow, 1940b: 9. — Type species: *Pseudamophilus malaisei* Bollow, 1940b. — Gender masculine.

davidi Kodada, 1992

Pseudamophilus davidi Kodada, 1992: 359

TYPE LOCALITY: Thailand.

DISTRIBUTION: Thailand.

japonicus Nomura, 1957a

Pseudamophilus japonicus Nomura, 1957a: 3

TYPE LOCALITY: Japan (Honshu).

DISTRIBUTION: Japan (Honshu, Kyushu).

malaisei Bollow, 1940b

Pseudamophilus malaisei Bollow, 1940b: 10

TYPE LOCALITY: Myanmar.

DISTRIBUTION: Myanmar.

tibiaferus Yang & Zhang, 2002

Pseudamophilus tibiaferus Yang & Zhang, 2002: 812

TYPE LOCALITY: China (Fujian).

DISTRIBUTION: China (Fujian).

NOTE: First described in a thesis by Zhang (1994: 54).

Pseudancyronyx Bertrand & Steffan, 1963

Pseudancyronyx Bertrand & Steffan, 1963: 829. — Type species: *Ancyronyx quadriguttatus* Delève, 1937a. — Gender masculine.

alluaudi (Grouvelle, 1911c)

Ancyronyx alluaudi Grouvelle, 1911c: 278

TYPE LOCALITY: Uganda.

DISTRIBUTION: Rwanda, Uganda.

basilewskyi (Janssens, 1962)

Ancyronyx basilewskyi Janssens, 1962: 431

TYPE LOCALITY: Tanzania.

DISTRIBUTION: Tanzania.

concolor (Grouvelle, 1920)

Ancyronyx concolor Grouvelle, 1920: 215

TYPE LOCALITY: Kenya.

DISTRIBUTION: Kenya.

humeralis humeralis (Grouvelle, 1906c)

Ancyronyx humeralis Grouvelle, 1906c: 328

TYPE LOCALITY: Tanzania.

DISTRIBUTION: Congo (DR), Kenya, Tanzania.

humeralis natalensis Delève, 1970e

Pseudancyronyx humeralis natalensis Delève, 1970e: 338

TYPE LOCALITY: South Africa.

DISTRIBUTION: South Africa, Swaziland.

insignis (Delève, 1938)

Ancyronyx insignis Delève, 1938: 374

TYPE LOCALITY: Congo (DR).

DISTRIBUTION: Congo (DR).

kivuensis Delève, 1970e

Pseudancyronyx kivuensis Delève, 1970e: 336
Type Locality: Congo (DR).
Distribution: Congo (DR).

posticalis (Delève, 1963f)

Ancyronyx posticalis Delève, 1963f: 823
Type Locality: Cameroon.
Distribution: Cameroon, Congo (R).

purpurascens (Grouvelle, 1920)

Ancyronyx purpurascens Grouvelle, 1920: 213
Type Locality: Kenya.
Distribution: Kenya.

quadriguttatus (Delève, 1937a)

Ancyronyx quadriguttatus Delève, 1937a: 163
Type Locality: Congo (DR).
Distribution: Congo (DR), Congo (R), Rwanda.

robustus (Delève, 1937a)

Ancyronyx robustus Delève, 1937a: 162
Type Locality: Congo (DR).
Distribution: Congo (DR).

spathifer (Delève, 1956)

Ancyronyx spathifer Delève, 1956: 381
Type Locality: Rwanda.
Distribution: Rwanda.

Pseudelmidolia Delève, 1963d

Pseudelmidolia Delève, 1963d: 9. — Type species: *Elmidolia biapicata* Fairmaire, 1898b. — Gender feminine.

atra Delève, 1963d

Pseudelmidolia atra Delève, 1963d: 38
Type Locality: Madagascar.
Distribution: Madagascar.

bertrandi Delève, 1963d

Pseudelmidolia bertrandi Delève, 1963d: 36

TYPE LOCALITY: Madagascar.

DISTRIBUTION: Madagascar.

biapicata (Fairmaire, 1898b)

Elmidolia biapicata Fairmaire, 1898b: 466

TYPE LOCALITY: Madagascar.

DISTRIBUTION: Madagascar.

colasi Delève, 1963d

Pseudelmidolia colasi Delève, 1963d: 43

TYPE LOCALITY: Madagascar.

DISTRIBUTION: Madagascar.

conspecta (Grouvelle, 1906a)

Elmidolia conspecta Grouvelle, 1906a: 156

TYPE LOCALITY: Madagascar.

DISTRIBUTION: Madagascar.

conspicua (Grouvelle, 1906a)

Elmidolia conspicua Grouvelle, 1906a: 157

TYPE LOCALITY: Madagascar.

DISTRIBUTION: Madagascar.

coriariocollis Delève, 1963d

Pseudelmidolia coriariocollis Delève, 1963d: 33

TYPE LOCALITY: Madagascar.

DISTRIBUTION: Madagascar.

crassa (Grouvelle, 1906a)

Elmidolia crassa Grouvelle, 1906a: 152

TYPE LOCALITY: Madagascar.

DISTRIBUTION: Madagascar.

disconcinna Delève, 1963d

Pseudelmidolia disconcinna Delève, 1963d: 24

TYPE LOCALITY: Madagascar.

DISTRIBUTION: Madagascar.

eximia Delève, 1963d

Pseudelmidolia eximia Delève, 1963d: 45

Type Locality: Madagascar.

Distribution: Madagascar.

fusca Delève, 1963d

Pseudelmidolia fusca Delève, 1963d: 32

Type Locality: Madagascar.

Distribution: Madagascar.

gregaria Delève, 1963d

Pseudelmidolia gregaria Delève, 1963d: 29

Type Locality: Madagascar.

Distribution: Madagascar.

humilis Delève, 1963d

Pseudelmidolia humilis Delève, 1963d: 19

Type Locality: Madagascar.

Distribution: Madagascar.

metatibialis Delève, 1963d

Pseudelmidolia metatibialis Delève, 1963d: 47

Type Locality: Madagascar.

Distribution: Madagascar.

minor (Fairmaire, 1898b)

Elmidolia minor Fairmaire, 1898b: 467

Type Locality: Madagascar.

Distribution: Madagascar.

nigricula Delève, 1963d

Pseudelmidolia nigricula Delève, 1963d: 23

Type Locality: Madagascar.

Distribution: Madagascar.

ochraceipennis (Grouvelle, 1899)

Helmis ochraceipennis Grouvelle, 1899: 183

Type Locality: Madagascar.

Distribution: Madagascar.

pallidipennis Delève, 1963d

Pseudelmidolia pallidipennis Delève, 1963d: 30

TYPE LOCALITY: Madagascar.

DISTRIBUTION: Madagascar.

pauliani Delève, 1963d

Pseudelmidolia pauliani Delève, 1963d: 44

TYPE LOCALITY: Madagascar.

DISTRIBUTION: Madagascar.

pinguis (Fairmaire, 1902)

Elmidolia pinguis Fairmaire, 1902: 343

TYPE LOCALITY: Madagascar.

DISTRIBUTION: Madagascar.

sordida (Grouvelle, 1906a)

Elmidolia sordida Grouvelle, 1906a: 151

TYPE LOCALITY: Madagascar.

DISTRIBUTION: Madagascar.

soror (Grouvelle, 1906a)

Elmidolia soror Grouvelle, 1906a: 150

TYPE LOCALITY: Madagascar.

DISTRIBUTION: Madagascar.

spreta Delève, 1963d

Pseudelmidolia spreta Delève, 1963d: 18

TYPE LOCALITY: Madagascar.

DISTRIBUTION: Madagascar.

starmuhleri Delève, 1963d

Pseudelmidolia starmuhleri Delève, 1963d: 35

TYPE LOCALITY: Madagascar.

DISTRIBUTION: Madagascar.

NOTE: Named after the Austrian scientist Ferdinand Starmühlner (1927–2006); the patronym ("Starmühlner") was incorrectly spelled as "Starmuhler" throughout the original description (altogether eight times!), therefore there is no evidence of an inadvertant error in the original publication itself and the epithet name cannot be changed to "*starmuehlneri*" (ICZN 1999: Art. 32.5.1).

striolata (Fairmaire, 1902)

Elmidolia striolata Fairmaire, 1902: 343 ("*Elmidola*")

TYPE LOCALITY: Madagascar.

DISTRIBUTION: Madagascar.

stulta (Grouvelle, 1906a)

Elmidolia stulta Grouvelle, 1906a: 152

TYPE LOCALITY: Madagascar.

DISTRIBUTION: Madagascar.

trinervosa Delève, 1963d

Pseudelmidolia trinervosa Delève, 1963d: 21

TYPE LOCALITY: Madagascar.

DISTRIBUTION: Madagascar.

umbrina (Fairmaire, 1898b)

Elmidolia umbrina Fairmaire, 1898b: 466

TYPE LOCALITY: Madagascar.

DISTRIBUTION: Madagascar.

verrucosa Delève, 1963d

Pseudelmidolia verrucosa Delève, 1963d: 20

TYPE LOCALITY: Madagascar.

DISTRIBUTION: Madagascar.

Pseudodisersus Brown, 1981c

Pseudodisersus Brown, 1981c: 98. — Type species: *Potamophilus goudotii* Guérin Méneville, 1843. — Gender masculine.

goudotii (Guérin Méneville, 1843)

Potamophilus goudotii Guérin Méneville, 1843: 18

TYPE LOCALITY: Colombia.

DISTRIBUTION: Colombia, Costa Rica, Ecuador, Panama.

SYNONYM: *Pseudodisersus coquereli* Brown, 1981c: 100. — Type locality: Colombia.

Pseudomacronychus Grouvelle, 1906c

Pseudomacronychus Grouvelle, 1906c: 326. — Type species: *Pseudomacronychus castaneus* Grouvelle, 1906c. — Gender masculine.

brevis Delève, 1965c

Pseudomacronychus brevis Delève, 1965c: 6

Type Locality: Congo (DR).

Distribution: Congo (DR), Congo (R).

castaneus castaneus Grouvelle, 1906c

Pseudomacronychus castaneus Grouvelle, 1906c: 326

Type Locality: Kenya.

Distribution: Angola, Congo (R), Kenya, Rwanda.

castaneus litigiosus Delève, 1945a

Pseudomacronychus castaneus litigiosus Delève, 1945a: 4

Type Locality: Congo (DR).

Distribution: Congo (DR).

castaneus miliaris Delève, 1965c

Pseudomacronychus castaneus miliaris Delève, 1965c: 4

Type Locality: Angola.

Distribution: Angola.

castaneus nitidus Delève, 1965c

Pseudomacronychus castaneus nitidus Delève, 1965c: 3

Type Locality: Angola.

Distribution: Angola.

decoratus Grouvelle, 1911c

Pseudomacronychus decoratus Grouvelle, 1911c: 277

Type Locality: Kenya.

Distribution: Congo (DR), Kenya, South Africa, Tanzania.

humeralis Delève, 1937a

Pseudomacronychus humeralis Delève, 1937a: 161

Type Locality: Congo (DR).

Distribution: Congo (DR), Congo (R), Rwanda.

intermedius Delève, 1963e

Pseudomacronychus intermedius Delève, 1963e: 90

Type Locality: Congo (DR).

Distribution: Congo (DR), Congo (R), Gabon, Ghana, Guinea, Ivory Coast.

lineatus Delève, 1967a

Pseudomacronychus lineatus Delève, 1967a: 340

Type Locality: Congo (R).

Distribution: Congo (R).

occidentalis Alluaud, 1933

Pseudomacronychus occidentalis Alluaud, 1933: 157

Type Locality: Ivory Coast.

Distribution: Burkina Faso, Congo (DR), Gabon, Ghana, Ivory Coast.

ovatus Delève, 1965c

Pseudomacronychus ovatus Delève, 1965c: 9

Type Locality: Congo (DR).

Distribution: Congo (DR).

rufilabris Delève, 1956

Pseudomacronychus rufilabris Delève, 1956: 379

Type Locality: Rwanda.

Distribution: Rwanda.

scutellatus circumcinctus Delève, 1968b

Pseudomacronychus scutellatus circumcinctus Delève, 1968b: 205

Type Locality: Ivory Coast.

Distribution: Ivory Coast.

scutellatus scutellatus Delève, 1965c

Pseudomacronychus scutellatus Delève, 1965c: 7

Type Locality: Congo (DR).

Distribution: Congo (DR), Gabon, Ghana, Liberia.

simulator Delève, 1945a

Pseudomacronychus simulator Delève, 1945a: 5

Type Locality: Congo (DR).

Distribution: Congo (DR), Congo (R).

variolosus Delève, 1945a

Pseudomacronychus variolosus Delève, 1945a: 8

Type Locality: Congo (DR).

Distribution: Congo (DR).

Rhizelmis Chandler, 1954

Rhizelmis Chandler, 1954: 126. — Type species: *Rhizelmis nigra* Chandler, 1954. — Gender feminine.

nigra Chandler, 1954

Rhizelmis nigra Chandler, 1954: 128

TYPE LOCALITY: USA (California).

DISTRIBUTION: USA (California).

Rhopalonychus Jäch & Kodada, 1996b

Rhopalonychus Jäch & Kodada, 1996b: 410. — Type species: *Rhopalonychus levatorponderis* Jäch & Kodada, 1996b. — Gender masculine.

levatorponderis Jäch & Kodada, 1996b

Rhopalonychus levatorponderis Jäch & Kodada, 1996b: 418

TYPE LOCALITY: Malaysia (Sarawak).

DISTRIBUTION: Malaysia (Sarawak).

Riolus Mulsant & Rey, 1872

Riolus Mulsant & Rey, 1872: 29. — Type species: *Limnius cupreus* Müller, 1806a. — Gender masculine.

SYNONYMS: *Normandia* Pic, 1900a: 267 **syn.n.** — Type species: *Limnius villosocostatus* Reiche, 1879. — Gender feminine.

Aptyktophallus Steffan, 1958: 165. — Type species: *Limnius nitens* Müller, 1817. — Gender masculine.

cupreus (Müller, 1806a)

Limnius cupreus Müller, 1806a: 205

TYPE LOCALITY: Germany.

DISTRIBUTION: Austria, Belgium, Bosnia and Herzegovina, Bulgaria, Croatia, Czech Republic, Denmark, Estonia, France, Germany, Greece, Hungary, Italy, Latvia, Lithuania, Luxembourg, (Montenegro), Netherlands, Poland, Romania, Russia, Serbia, Slovakia, Slovenia, Spain, Sweden, Switzerland, Turkey, Ukraine, United Kingdom.

SYNONYMS: *Riolus steineri* Kuwert, 1889: 24. — Type locality: Spain. NOTE: Originally described in a key; formal description published by Kuwert (1890: 51).

Riolus steineri senex Kuwert, 1889: 24. — Type locality: Italy. NOTE: Originally described as a variety of *Riolus steineri* Kuwert, 1889 (= *R. cupreus* (Müller, 1806a)) in a key; formal description (still as variety) published by Kuwert (1890: 51).

Riolus erichsoni Kuwert, 1889: 25. — Type locality: Central Europe ("Mitteleuropa"), according to original description (Kuwert 1889: 25) and Kuwert (1890: 33). NOTE: Specimens regarded by Berthélemy (1979: 67) as "syntypes" and "paratypes" are from Bosnia and Herzegovina ("Bosnie") and Montenegro ("Petrovač") and therefore cannot be regarded as type specimens.

Riolus mulsanti Kuwert, 1889: 25. — Type locality: Switzerland, according to lectotype designation by Berthélemy (1979: 68); however, in the original description only "England, Normandie, Holland" are mentioned, therefore, it will have to be checked whether the specimen designated by Berthélemy (1979) as lectotype is really a syntype.

Riolus lentzi Kuwert, 1889: 25. — Type locality: France. NOTE: Originally described in a key; formal description published by Kuwert (1890: 52).

Riolus argolicus Janssens, 1959b: 16. — Type locality: Greece.

fontinalis Jäch, 1984a

Riolus fontinalis Jäch, 1984a: 140

TYPE LOCALITY: Turkey.

DISTRIBUTION: Turkey.

illiesi Steffan, 1958

Riolus illiesi Steffan, 1958: 159

TYPE LOCALITY: Germany.

DISTRIBUTION: Belgium, France, Germany, Spain, Switzerland.

nitens (Müller, 1817)

Limnius nitens Müller, 1817: 273

TYPE LOCALITY: Germany.

DISTRIBUTION: Algeria, Austria, Belarus, Belgium, Bosnia and Herzegovina, Croatia, Estonia, Finland, France, Germany, Greece, Israel, Italy, Latvia, Lebanon, Lithuania, Macedonia, Morocco, Netherlands, Poland? (Przewoźny et al. 2011), Portugal, Russia, Slovenia, Spain, Sweden, Switzerland, Turkey, Ukraine, United Kingdom.

SYNONYMS: *Elmis orichalceus* Heer, 1841: 470. — Type locality: Switzerland. NOTE: The name was first published as a nomen nudum ("*L.* [*Limnius*] *orichalceus*") by Gyllenhal (1827: 395) under "*L.* [*Limnius*] *cupreus*" (= *Riolus cupreus* (Müller, 1806a)); listed as a junior synonym of *Riolus nitens* (Müller, 1817) by Zaitzev (1908: 307, 1910: 33) and Mascagni & Calamandrei (1992: 132).

Riolus sauteri Kuwert, 1889: 24. — Type locality: France (Corsica). NOTE: Originally described in a key; formal description published by Kuwert (1890: 50).

Riolus seidlitzi Kuwert, 1889: 24. — Type locality: France. NOTE: Originally described in a key; formal description published by Kuwert (1890: 50).

robustior Pic, 1900a

Riolus villosocostatus robustior Pic, 1900a: 266

TYPE LOCALITY: Tunisia.

DISTRIBUTION: Algeria, Tunisia.

NOTE: Originally described as a variety of *Riolus villosocostatus* (Reiche, 1879).

sodalis (Erichson, 1847)

Elmis sodalis Erichson, 1847: 532

TYPE LOCALITY: Germany.

DISTRIBUTION: France, Germany, Italy, Spain, Switzerland.

SYNONYM: *Riolus meridionalis* Grouvelle, 1900c: 137. — Type locality: France.

somcheticus (Kolenati, 1846)

Elmis somchetica Kolenati, 1846: 64

TYPE LOCALITY: Georgia.

DISTRIBUTION: Armenia, Georgia, Greece (Samos), Iran, Lebanon, Russia, Syria, Turkey, Ukraine.

NOTE: Incorrect original spelling: "*Elmis somcheticus*"; incorrect subsequent spellings: "*soncheticus*" (Lacordaire 1854: 509), "*som(c)heticus*" (Zaitzev 1908: 307, 1910: 33); transferred to *Riolus* by Kuwert (1889: 26); "redescribed" and transferred to *Elmis* by Zaitzev (1947: 89), but redescription not based on type material; a female syntype (labelled: "Tiflis", "Kolenati") deposited in the Naturhistorisches Museum Wien, Austria, was examined by the first author (M.A. Jäch).

SYNONYMS: *Elmis syriaca* Allard, 1869: 466 **syn.n.** — Type locality: Lebanon. NOTE: Incorrect original spelling: "*Elmis syriacus*".

Riolus anatolicus Delève, 1963b: 9. — Type locality: Turkey.

substriatus Grouvelle, 1889a

Riolus substriatus Grouvelle, 1889a: LXXX[80]

TYPE LOCALITY: Algeria.

DISTRIBUTION: Algeria, Morocco.

SYNONYM: *Riolus substriatus punctatostriatus* Alluaud, 1922: 38. — Type locality: Morocco. NOTE: Originally described as a variety of *Riolus substriatus.*

subviolaceus (Müller, 1817)

Limnius subviolaceus Müller, 1817: 273

TYPE LOCALITY: Probably in Germany; according to the original description (p. 273, footnote) the type material was collected by Christian Gottfried Daniel Nees von Esenbeck (1776–1858), a prolific German botanist, physician, zoologist, and natural philosopher, who described approximately 7,000 plant species (almost as many as Linnæus).

DISTRIBUTION: Austria, Belgium, Bosnia and Herzegovina, Bulgaria, Croatia, Czech Republic, France, Germany, Greece, Hungary, Italy, Luxembourg, (Montenegro), Netherlands, Poland, Romania, Serbia, Slovakia, Slovenia, Spain, Switzerland, United Kingdom.

NOTE: Incorrect subsequent spelling: "*sub-violaceus*" (Dejean 1821: 49).

SYNONYMS: *Riolus subviolaceus auronitens* Kuwert, 1889: 26. — Type locality: Not clearly identifiable; according to the original description: "Krain [Slovenia], Bosnien [Bosnia and Herzegovina]"; according to lectotype designation by Berthélemy (1979: 69): "Hungarn" [Hungary] (in the late 19th century the territory of Hungary included Croatia (without Dalmatia), Slovakia, large parts of Romania, parts of Serbia and smaller parts of Poland and the Ukraine); it will have to be checked whether the specimen designated by Berthélemy (1979) as lectotype is really a syntype. NOTE: Originally described as a variety of *Riolus subviolaceus* in a key.

Riolus subviolaceus bosnicus Kuwert, 1889: 26. — Type locality: Croatia or Bosnia and Herzegovina; according to the original description: "Slavonien [eastern Croatia], Bosnien [Bosnia and Herzegovina]"; Berthélemy (1979: 69) stated that the types of *Riolus subviolaceus bosnicus* originate from Bosnia and Herzegovina ("Petrovač et de Sarajevo"), but this is not based on lectotype designation. NOTE: Originally described as a variety of *Riolus subviolaceus* in a key.

Riolus australis Rey, 1889: 67. — Type locality: France.

Riolus nigricans Rey, 1889: 67. — Type locality: France.

Riolus apfelbecki Ganglbauer, 1904b: 354. — Type locality: Serbia.

Riolus wichmanni Zimmermann, 1908: 341. — Type locality: Austria.

villosocostatus (Reiche, 1879)

Limnius villosocostatus Reiche, 1879: 238

TYPE LOCALITY: Algeria.

DISTRIBUTION: Algeria, Morocco, Tunisia.

NOTE: Incorrect original spelling: "*villoso-costatus*"; transferred to *Riolus* by Grouvelle (1889a: LXXX[80]).

Roraima Kodada & Jäch, 1999

Roraima Kodada & Jäch, 1999: 14. — Type species: *Roraima carinata* Kodada & Jäch, 1999. — Gender feminine.

carinata Kodada & Jäch, 1999

Roraima carinata Kodada & Jäch, 1999: 21

TYPE LOCALITY: Venezuela.

DISTRIBUTION: Venezuela.

Rudielmis Jäch & Boukal, 1995b

Rudielmis Jäch & Boukal, 1995b: 151. — Type species: *Rudielmis schuhi* Jäch & Boukal, 1995b. — Gender feminine.

concolor Jäch & Boukal, 1995b

Rudielmis concolor Jäch & Boukal, 1995b: 156

TYPE LOCALITY: India (Kerala).

DISTRIBUTION: India (Kerala).

schuhi Jäch & Boukal, 1995b

Rudielmis schuhi Jäch & Boukal, 1995b: 152

TYPE LOCALITY: India (Kerala).

DISTRIBUTION: India (Karnataka, Kerala).

Simsonia Carter & Zeck, 1929

Simsonia Carter & Zeck, 1929: 58. — Type species: *Elmis tasmanica* Blackburn, 1894. — Gender feminine.

allmani Carter & Zeck, 1936

Simsonia allmani Carter & Zeck, 1936: 158

TYPE LOCALITY: Australia (New South Wales).

DISTRIBUTION: Australia (New South Wales).

ambigua Jäch, 1985

Simsonia ambigua Jäch, 1985: 242

TYPE LOCALITY: Papua New Guinea.

DISTRIBUTION: Papua New Guinea.

angusta (Carter, 1926b)

Helmis angusta Carter, 1926b: 62

TYPE LOCALITY: Australia (Victoria).

DISTRIBUTION: Australia (Victoria).

bispina Jäch, 1985

Simsonia bispina Jäch, 1985: 242

TYPE LOCALITY: Papua New Guinea.

DISTRIBUTION: Papua New Guinea.

brooksi Carter & Zeck, 1938

Simsonia brooksi Carter & Zeck, 1938: 170

TYPE LOCALITY: Australia (Queensland).

DISTRIBUTION: Australia (Queensland).

cotterensis Carter & Zeck, 1933

Simsonia cotterensis Carter & Zeck, 1933: 371

TYPE LOCALITY: Australia (ACT).

DISTRIBUTION: Australia (ACT).

deanei Carter, 1930

Simsonia deanei Carter, 1930: 189

TYPE LOCALITY: Australia (Queensland).

DISTRIBUTION: Australia (Queensland).

eborica Carter & Zeck, 1935

Simsonia eborica Carter & Zeck, 1935: 79

TYPE LOCALITY: Australia (New South Wales).

DISTRIBUTION: Australia (New South Wales).

NOTE: In the original description this species was spelled as "*eborica*" (1 ×) and as "*eborensis*" (1 ×); Carter & Zeck (1936: 159) selected "*eborica*" as the correct spelling (ICZN 1999: Art. 24.2.3).

hopsoni Carter & Zeck, 1929

Simsonia hopsoni Carter & Zeck, 1929: 58

TYPE LOCALITY: Australia (New South Wales).

DISTRIBUTION: Australia (New South Wales).

NOTE: Might belong to *Notriolus* Carter & Zeck, 1929 (Glaister 1999: 34).

irregularis Carter & Zeck, 1929

Simsonia irregularis Carter & Zeck, 1929: 60

TYPE LOCALITY: Australia (New South Wales).

DISTRIBUTION: Australia (New South Wales).

NOTE: Might belong to *Notriolus* Carter & Zeck, 1929 (Glaister 1999: 34).

leai Carter & Zeck, 1929

Simsonia leai Carter & Zeck, 1929: 59

TYPE LOCALITY: Australia (New South Wales).

DISTRIBUTION: Australia (New South Wales, Tasmania).

longipes Carter & Zeck, 1933

Simsonia longipes Carter & Zeck, 1933: 371

TYPE LOCALITY: Australia (Queensland).

DISTRIBUTION: Australia (Queensland, Victoria).

neoguineana Satô, 1973

Simsonia neoguineana Satô, 1973: 466

TYPE LOCALITY: Papua New Guinea.

DISTRIBUTION: Papua New Guinea.

nicholsoni (Carter, 1926b)

Helmis nicholsoni Carter, 1926b: 61

TYPE LOCALITY: Australia (New South Wales).

DISTRIBUTION: Australia (New South Wales, Victoria).

NOTE: Might belong to *Notriolus* Carter & Zeck, 1929 (Glaister 1999: 34).

SYNONYM: *Helmis nicholsoni bicolor* Carter, 1926b: 61. — Type locality: Australia (Victoria).
NOTE: Originally described as "var. *H. bicolor*" under *Helmis nicholsoni*; might be a good species.

purpurea (Carter, 1926a)

Helmis purpurea Carter, 1926a: 508

TYPE LOCALITY: Australia (New South Wales).

DISTRIBUTION: Australia (New South Wales).

quadrimaculata Jäch, 1985

Simsonia quadrimaculata Jäch, 1985: 244

TYPE LOCALITY: Papua New Guinea.

DISTRIBUTION: Papua New Guinea.

tasmanica (Blackburn, 1894)

Elmis tasmanica Blackburn, 1894: 94

TYPE LOCALITY: Australia (Tasmania).

DISTRIBUTION: Australia (Tasmania).

NOTE: Incorrect original spelling: "*Elmis tasmanicus*"; might belong to *Notriolus* Carter & Zeck, 1929 (Glaister 1999: 34).

tonnoiri Carter & Zeck, 1936

Simsonia tonnoiri Carter & Zeck, 1936: 158

TYPE LOCALITY: Australia (New South Wales).

DISTRIBUTION: Australia (ACT, New South Wales, Victoria).

vestita Carter & Zeck, 1929

Simsonia vestita Carter & Zeck, 1929: 60

Type Locality: Australia (Victoria).

Distribution: Australia (Victoria).

Note: Might belong to *Notriolus* Carter & Zeck, 1929 (Glaister 1999: 34).

wilsoni (Carter, 1926b)

Helmis wilsoni Carter, 1926b: 64

Type Locality: Australia (Victoria).

Distribution: Australia (Victoria).

Sinelmis Satô & Kishimoto, 2001

Sinelmis Satô & Kishimoto, 2001: 76. — Type species: *Sinelmis uenoi* Satô & Kishimoto, 2001. — Gender feminine.

Note: This genus might be a junior synonym of *Ordobrevia* Sanderson, 1953a or *Stenelmis* Dufour, 1835.

uenoi Satô & Kishimoto, 2001

Sinelmis uenoi Satô & Kishimoto, 2001: 77

Type Locality: China (Guizhou).

Distribution: China (Guizhou).

Sinonychus Jäch & Boukal, 1995a

Sinonychus Jäch & Boukal, 1995a: 306. — Type species: *Sinonychus lantau* Jäch & Boukal, 1995a. — Gender masculine.

lantau Jäch & Boukal, 1995a

Sinonychus lantau Jäch & Boukal, 1995a: 308

Type Locality: China (Hong Kong).

Distribution: China (Hong Kong).

satoi Yoshitomi & Nakajima, 2007

Sinonychus satoi Yoshitomi & Nakajima, 2007: 97

Type Locality: Japan (Ryukyu Islands).

Distribution: Japan (Ryukyu Islands).

tsujunensis Yoshitomi & Nakajima, 2012

Sinonychus tsujunensis Yoshitomi & Nakajima, 2012: 55

Type Locality: Japan (Kyushu).

Distribution: Japan (Kyushu).

Sphragidelmis Delève, 1964a

Sphragidelmis Delève, 1964a: 26. — Type species: *Limnius ikopae* Fairmaire, 1898a. — Gender feminine.

atomaria (Fairmaire, 1898b)

Limnius atomarius Fairmaire, 1898b: 467

Type Locality: Madagascar.

Distribution: Madagascar.

Note: Incorrect subsequent spelling: "*Sphragidelmis atomarius*" (Delève 1964a: 29).

bothrideres (Fairmaire, 1902)

Limnius bothrideres Fairmaire, 1902: 345

Type Locality: Madagascar.

Distribution: Madagascar.

Note: Possibly a junior synonym of *Sphragidelmis ikopae* (Fairmaire, 1898a) (see Delève 1964a: 30).

ikopae (Fairmaire, 1898a)

Limnius ikopae Fairmaire, 1898a: 225

Type Locality: Madagascar.

Distribution: Madagascar.

Note: Incorrect original spelling: "*ikopœ*".

trilineata (Grouvelle, 1906a)

Limnius trilineatus Grouvelle, 1906a: 148

Type Locality: Madagascar.

Distribution: Madagascar.

Note: Incorrect subsequent spelling: "*Sphragidelmis trilineatus*" (Delève 1964a: 29).

Stegoelmis Hinton, 1939e

Stegoelmis Hinton, 1939e: 30. — Type species: *Stenelmis geayi* Grouvelle, 1908. — Gender feminine.

andersoni Spangler, 1990

Stegoelmis andersoni Spangler, 1990: 10

Type Locality: Ecuador.

Distribution: Brazil (Rondônia) (Manzo 2013: 212), Colombia, Ecuador, Paraguay, Peru.

crinita Spangler, 1990

Stegoelmis crinita Spangler, 1990: 12

Type Locality: Ecuador.

Distribution: Ecuador.

ennsi Spangler, 1990

Stegoelmis ennsi Spangler, 1990: 36

Type Locality: Venezuela.

Distribution: Venezuela.

fera Spangler, 1990

Stegoelmis fera Spangler, 1990: 33

Type Locality: Venezuela.

Distribution: Venezuela.

figueiredoensis Fernandes, Passos & Hamada, 2011

Stegoelmis figueiredoensis Fernandes, Passos & Hamada, 2011: 58

Type Locality: Brazil (Amazonas).

Distribution: Brazil (Amazonas).

geayi (Grouvelle, 1908)

Stenelmis geayi Grouvelle, 1908: 181

Type Locality: French Guiana.

Distribution: Ecuador, French Guiana, Guyana, Venezuela.

Synonym: *Stegoelmis hintoni* Sanderson, 1953b: 34. — Type locality: Ecuador.

ica Spangler, 1990

Stegoelmis ica Spangler, 1990: 40

Type Locality: Brazil (Amazonas).

Distribution: Brazil (Amazonas).

selva Spangler, 1990

Stegoelmis selva Spangler, 1990: 15

Type Locality: Venezuela.

Distribution: Venezuela.

shepardi Fernandes, Passos & Hamada, 2011

Stegoelmis shepardi Fernandes, Passos & Hamada, 2011: 61

Type Locality: Brazil (Roraima).

Distribution: Brazil (Roraima), French Guiana.

sticta Spangler, 1990

Stegoelmis sticta Spangler, 1990: 49

Type Locality: Ecuador.

Distribution: Colombia, Ecuador.

stictoides Spangler, 1990

Stegoelmis stictoides Spangler, 1990: 39

Type Locality: Venezuela.

Distribution: Brazil (Amazonas), Guyana, Suriname, Venezuela.

tuberosa Spangler, 1990

Stegoelmis tuberosa Spangler, 1990: 28

Type Locality: Venezuela.

Distribution: Venezuela.

verrucata Hinton, 1939e

Stegoelmis verrucata Hinton, 1939e: 31

Type Locality: Brazil (Pará).

Distribution: Brazil (Pará), French Guiana.

Stenelmis Dufour, 1835

Stenelmis Dufour, 1835: 158. — Type species: *Limnius canaliculatus* Gyllenhal, 1808. — Gender feminine.

Note: Incorrect subsequent spelling: "*Stenhelmis*" Zaitzev (1908: 299), not to be regarded as an emendation (ICZN 1999: Art. 33.2.1).

accessoria Yang & Zhang, 2002

Stenelmis accessoria Yang & Zhang, 2002: 818

Type Locality: China (Fujian).

Distribution: China (Anhui, Fujian, Guangxi).

adusta Delève, 1938

Stenelmis adusta Delève, 1938: 356

Type Locality: Congo (DR).

Distribution: Angola, Congo (DR).

alluaudi Grouvelle, 1906c

Stenelmis alluaudi Grouvelle, 1906c: 317

Type Locality: Tanzania.

Distribution: Tanzania.

aloysiisabaudiae Pic, 1930

Stenelmis aloysiisabaudiae Pic, 1930: 23

Type Locality: Ethiopia; the "type" of this species was collected in January 1929 by Luigi Amedeo Giuseppe Maria Ferdinando Francesco di Savoia, Duke of the Abruzzi (1873–1933) during his expedition to the source of Shebele River; although the original description refers to "Somalie Italienne" (Somalia), the type locality ("Malca-Dubbá", Imi Region) belongs to Ethiopia.

Distribution: Ethiopia.

Note: Incorrect original spelling: "*aloysii-sabaudiae*"; incorrect subsequent spellings: "*aloyssi - sabaudiae*" (Delève 1966a: 57), "*aloyssii sabaudiae*" (Delève 1966a: 61), "*aloyssii-sabaudiae*" (Delève 1966a: 96).

ampliata Delève, 1938

Stenelmis ampliata Delève, 1938: 355

Type Locality: Congo (DR).

Distribution: Congo (DR), Congo (R).

anderssoni Delève, 1973a

Stenelmis anderssoni Delève, 1973a: 9

Type Locality: Sri Lanka.

Distribution: Sri Lanka.

angolensis Delève, 1966d

Stenelmis angolensis Delève, 1966d: 54

Type Locality: Angola.

Distribution: Angola, Central African Republic.

angustisulcata Zhang & Yang, 1995

Stenelmis angustisulcata Zhang & Yang, 1995: 103

Type Locality: China (Zhejiang).

Distribution: China (Zhejiang).

antennalis Sanderson, 1938a

Stenelmis antennalis Sanderson, 1938a: 695

Type Locality: USA (Mississippi).

Distribution: USA (Alabama, Florida, Georgia, Mississippi, New Jersey, New York, North Carolina, South Carolina, Texas, Wisconsin).

anytus Hinton, 1941b

Stenelmis anytus Hinton, 1941b: 82

TYPE LOCALITY: China (Fujian).

DISTRIBUTION: China (Fujian).

NOTE: Epithet name is a noun in apposition (ancient Athenian politician).

aphela Alluaud, 1933

Stenelmis aphela Alluaud, 1933: 156

TYPE LOCALITY: Ivory Coast.

DISTRIBUTION: Burkina Faso, Ghana, Guinea, Ivory Coast, Liberia.

ares Hinton, 1941b

Stenelmis ares Hinton, 1941b: 92

TYPE LOCALITY: South Africa.

DISTRIBUTION: South Africa.

aria Janssens, 1961

Stenelmis aria Janssens, 1961: 5

TYPE LOCALITY: Afghanistan.

DISTRIBUTION: Afghanistan.

aritai Satô, 1964b

Stenelmis aritai Satô, 1964b: 32

TYPE LOCALITY: Japan (Ryukyu Islands).

DISTRIBUTION: Japan (Ryukyu Islands).

auriculata Yang & Zhang, 2002

Stenelmis auriculata Yang & Zhang, 2002: 820

TYPE LOCALITY: China (Fujian).

DISTRIBUTION: China (Fujian, Guizhou).

beameri Sanderson, 1938a

Stenelmis beameri Sanderson, 1938a: 671

TYPE LOCALITY: USA (Arkansas).

DISTRIBUTION: USA (Arkansas, Missouri, Oklahoma, Texas).

beijingana Zhang & Ding, 1995

Stenelmis beijingana Zhang & Ding, 1995: 15

TYPE LOCALITY: China (Beijing).

DISTRIBUTION: China (Beijing).

bicarinata Le Conte, 1852

Stenelmis bicarinata Le Conte, 1852: 44

TYPE LOCALITY: USA (Ohio).

DISTRIBUTION: Canada (Manitoba, Nova Scotia, Ontario, Quebec, Saskatchewan), Mexico, USA (Florida, Kansas, Mississippi, New Jersey, New Mexico, New York, Ohio, Oklahoma, Pennsylvania, Tennessee, Texas, South Carolina, Wisconsin).

NOTE: Incorrect original spelling: "*bicarinatus*".

SYNONYMS: *Stenelmis maerkelii* Motschulsky, 1854: 12. — Type locality: USA: (Tennessee), see Sanderson (1938a: 706). NOTE: Incorrect original spelling: "*märkelii*"; named after German entomologist Johann Christian Friedrich Märkel (1790–1860); incorrect subsequent spelling: "*maerkeli*" (Zaitzev 1908: 300, 1910: 23).

Stenelmis vittipennis Zimmermann, 1869: 259. — Type locality: USA (South Carolina). NOTE: The author of this species was spelled incorrectly ("Zimmerman") by Brown (1972a: 21).

Stenelmis convexula Sanderson, 1938a: 704. — Type locality: USA (Florida).

bicolor Reitter, 1886

Stenelmis bicolor Reitter, 1886: 213

TYPE LOCALITY: Indonesia (Sumatra).

DISTRIBUTION: Indonesia (Sumatra).

birmanica Grouvelle, 1892b

Stenelmis birmanica Grouvelle, 1892b: 867

TYPE LOCALITY: Myanmar.

DISTRIBUTION: Myanmar.

bosschae Grouvelle, 1892a

bosschae Grouvelle, 1892a: 188

TYPE LOCALITY: Indonesia (Borneo).

DISTRIBUTION: Indonesia (Borneo).

NOTE: Incorrect subsequent spelling: "*boschai*" (Zaitzev 1908: 299, 1910: 22).

brevipes Delève, 1968c

Stenelmis brevipes Delève, 1968c: 156

TYPE LOCALITY: Vietnam.

DISTRIBUTION: Vietnam.

brincki Delève, 1973a

Stenelmis brincki Delève, 1973a: 8

TYPE LOCALITY: Sri Lanka.

DISTRIBUTION: Sri Lanka.

calceata Delève, 1963f

Stenelmis calceata Delève, 1963f: 808

Type Locality: Cameroon.

Distribution: Cameroon.

calida Chandler, 1949

Stenelmis calida Chandler, 1949: 133

Type Locality: USA (Nevada).

Distribution: USA (Nevada).

Note: Incorrect subsequent spelling: "*calidae*" (Minckley & Deacon 1975: 107, 108).

canaliculata (Gyllenhal, 1808)

Limnius canaliculatus Gyllenhal, 1808: 552

Type Locality: Sweden.

Distribution: Austria, Belgium, Croatia, Finland, France, Germany, Italy, Luxembourg, Netherlands, Norway, Poland, Portugal, Slovakia, Spain, Sweden, Switzerland, Ukraine, United Kingdom.

Note: Incorrect subsequent spelling: "*Stenelmis canaliculatus*" (Erichson 1847: 534, Grenier 1863: 34, Bertolini 1874: 102, 1900: 26, Tamutis et al. 2011: 214).

Synonyms: *Stenelmis elongata* Motschulsky, 1860: 51. — Type locality: "Il m'a été donné comme venant de l'Amérique septentrionale, sans indication plus précise" (Motschulsky 1860: 51), "Am.[erica] bor.[ealis] ?" (Sanderson 1938a: 711); obviously, the holotype of this species is of European origin and has been wrongly labelled. Note: Incorrect original spelling: "*elongatus*"; synonymized by Sanderson (1938a: 711).

Stenelmis stoeckleini Bollow, 1941: 12. — Type locality: Switzerland.

Stenelmis bollovotiva Steffan, 1961: 307. — Type locality: Germany.

capys Hinton, 1941b

Stenelmis capys Hinton, 1941b: 78

Type Locality: China (Fujian).

Distribution: China (Fujian).

carbonaria Delève, 1966a

Stenelmis carbonaria Delève, 1966a: 75

Type Locality: Congo (DR).

Distribution: Congo (DR).

cardoni Grouvelle, 1894

Stenelmis cardoni Grouvelle, 1894: 586

TYPE LOCALITY: India (Jharkhand).

DISTRIBUTION: India (Jharkhand).

casius Hinton, 1941b

Stenelmis casius Hinton, 1941b: 80

TYPE LOCALITY: China (Fujian).

DISTRIBUTION: China (Fujian).

NOTE: Epithet name obviously is a noun in apposition; incorrect subsequent spelling: "*casiuns*" (Hua 2002: 98).

cavicula Delève, 1966a

Stenelmis cavicula Delève, 1966a: 77

TYPE LOCALITY: Tanzania ("Kwamkugo" [= Kwamkuyo]).

DISTRIBUTION: Tanzania.

centridivisa Zhang, Yang & Zhang, 2003

Stenelmis centridivisa Zhang, Yang & Zhang, 2003: 121

TYPE LOCALITY: China (Guangxi).

DISTRIBUTION: China (Guangxi).

chappuisi Alluaud, 1933

Stenelmis chappuisi Alluaud, 1933: 155

TYPE LOCALITY: Ivory Coast.

DISTRIBUTION: Angola, Burkina Faso, Guinea, Ivory Coast.

cheryl Brown, 1987

Stenelmis cheryl Brown, 1987: 111

TYPE LOCALITY: USA (Texas).

DISTRIBUTION: Canada (Quebec), Mexico, USA (Iowa, Kansas, Maine, Michigan, Minnesota, Missouri, New Hampshire, New Mexico, New York, Oklahoma, Pennsylvania, Texas, Vermont, West Virginia, Wisconsin).

clavareaui Grouvelle, 1900a

Stenelmis clavareaui Grouvelle, 1900a: 425

TYPE LOCALITY: Congo (DR).

DISTRIBUTION: Angola, Burkina Faso, Central African Republic, Congo (DR), Congo (R), Gabon, Guinea, Liberia, Rwanda.

SYNONYM: *Stenelmis carinata* Pic, 1923: 4. — Type locality: Gabon. NOTE: Synonymized by Delève (1966a: 81).

collaris Bollow, 1940b

Stenelmis collaris Bollow, 1940b: 13

TYPE LOCALITY: Myanmar.

DISTRIBUTION: Myanmar.

concinna Sanderson, 1938a

Stenelmis concinna Sanderson, 1938a: 674

TYPE LOCALITY: USA (New York).

DISTRIBUTION: Canada (New Brunswick, Quebec), USA (Alabama, Connecticut, Maine, Maryland, Massachusetts, New Hampshire, New York, North Carolina, Rhode Island, South Carolina, Tennessee, Vermont, Virginia, West Virginia).

concolor Grouvelle, 1896d

Stenelmis concolor Grouvelle, 1896d: 48

TYPE LOCALITY: Indonesia (Sumatra).

DISTRIBUTION: Indonesia (Sumatra).

confusa Delève, 1966a

Stenelmis confusa Delève, 1966a: 94

TYPE LOCALITY: Burkina Faso.

DISTRIBUTION: Burkina Faso, Ivory Coast.

consobrina consobrina Dufour, 1835

Stenelmis consobrina Dufour, 1835: 161

TYPE LOCALITY: France.

DISTRIBUTION: Algeria (Gauthier 1928: 39), Armenia, Bosnia and Herzegovina, Croatia (Mičetić Stanković et al. 2015: 105), Czech Republic, France, Georgia, Germany, Greece, Israel, Italy, Morocco, Palestine (West Bank), Russia, Spain, Switzerland, Syria, Tunisia (Boumaiza 1994: 213), Turkey, Turkmenistan.

NOTE: Incorrect subsequent spellings: "*consobrinus*" (Grenier 1863: 34, Bertolini 1874: 102, 1900: 26), "*consobria*" (Zhang & Yang 2003: 276).

consobrina peropaca Reitter, 1907

Stenelmis peropaca Reitter, 1907: 483

TYPE LOCALITY: Kazakhstan.

DISTRIBUTION: Afghanistan, China (Xinjiang), Kazakhstan, Uzbekistan.

constricta (Reitter, 1886)

Ancyronyx constrictus Reitter, 1886: 214

TYPE LOCALITY: Indonesia (Sumatra).

DISTRIBUTION: Indonesia (Sumatra).
NOTE: Transferred to *Stenelmis* by Jäch (1994a: 602).

convexa Zhang, Yang & Zhang, 2003

Stenelmis convexa Zhang, Yang & Zhang, 2003: 120
TYPE LOCALITY: China (Guangxi).
DISTRIBUTION: China (Guangxi).

corpulenta Delève, 1968c

Stenelmis corpulenta Delève, 1968c: 162
TYPE LOCALITY: Vietnam.
DISTRIBUTION: Vietnam.

crenata (Say, 1824)

Elmis crenata Say, 1824: 275
TYPE LOCALITY: USA (Tennessee), based on neotype designation (Sanderson 1938a: 667).
DISTRIBUTION: Canada (Manitoba, New Brunswick, Newfoundland, Nova Scotia, Ontario, Prince Edward Island, Quebec), USA (Alabama, Arkansas, Colorado, Florida, Georgia, Illinois, Indiana, Iowa, Kentucky, Louisiana, Maine, Massachusetts, Michigan, Minnesota, Mississippi, Missouri, Nebraska, New Hampshire, New York, North Carolina, Ohio, Oklahoma, Pennsylvania, Rhode Island, South Carolina, South Dakota, Tennessee, Texas, Vermont, Virginia, West Virginia).
NOTE: Incorrect original spelling: "*crenatus*".
SYNONYM: *Stenelmis sordida* Motschulsky, 1860: 51. — Type locality: USA (Pennsylvania).
NOTE: Incorrect original spelling: "*sordidus*".

cristata Delève, 1966a

Stenelmis cristata Delève, 1966a: 63
TYPE LOCALITY: Ivory Coast.
DISTRIBUTION: Cameroon, Congo (R), Ghana, Guinea, Ivory Coast, Liberia.

debilis Champion, 1927

Stenelmis debilis Champion, 1927: 51
TYPE LOCALITY: India (Himachal Pradesh).
DISTRIBUTION: India (Himachal Pradesh).

decellei Delève, 1966a

Stenelmis decellei Delève, 1966a: 85
TYPE LOCALITY: Ivory Coast.
DISTRIBUTION: Ghana, Ivory Coast, Liberia.

decipiens Bollow, 1941

Stenelmis decipiens Bollow, 1941: 39

Type Locality: China (Fujian).

Distribution: China (Fujian).

decorata Sanderson, 1938a

Stenelmis decorata Sanderson, 1938a: 701

Type Locality: USA (Kansas).

Distribution: USA (Alabama, Arkansas, Florida, Georgia, Illinois, Indiana, Iowa, Kansas, Kentucky, Louisiana, Michigan, Mississippi, Missouri, Oklahoma, South Carolina, Texas, Wisconsin).

dispar Delève, 1968c

Stenelmis dispar Delève, 1968c: 164

Type Locality: Vietnam.

Distribution: Vietnam.

dolon Hinton, 1941b

Stenelmis dolon Hinton, 1941b: 89

Type Locality: Malaysia (Perak).

Distribution: Malaysia (Perak).

douglasensis Sanderson, 1938a

Stenelmis douglasensis Sanderson, 1938a: 685

Type Locality: USA (Michigan).

Distribution: USA (Illinois, Indiana, Michigan, Wisconsin).

elfriedeae Bollow, 1941

Stenelmis elfriedeae Bollow, 1941: 64

Type Locality: China (Guangdong).

Distribution: China (Guangdong).

Note: Incorrect subsequent spelling: "*flfridae*" (Zhang & Yang 2003: 275).

euronotana Yang & Zhang, 1995

Stenelmis euronotana Yang & Zhang, 1995: 102

Type Locality: China (Fujian).

Distribution: China (Fujian, Guangxi, Zhejiang).

Note: Originally described in a key by Zhang & Yang (1995), authorship ascribed to Yang et Zhang (Zhang & Yang 1995: 104, 109); partly misspelled in original description: "*euronotara*"

(p. 109); formal description published twice, by Yang & Zhang (2002: 814) and (under reversed authorship) by Zhang & Yang (2003: 277); authorship erroneously ascribed to "Zhang & Yang" by Jäch et al. (2006: 438).

exarata Motschulsky, 1860

Stenelmis exarata Motschulsky, 1860: 49

Type Locality: "Continent Indian" (Motschulsky 1860: 49).

Distribution: India?

Note: Incorrect original spelling: "*exaratus*".

exigua Sanderson, 1938a

Stenelmis exigua Sanderson, 1938a: 669

Type Locality: USA (Arkansas).

Distribution: USA (Arkansas, Missouri, Oklahoma, Texas).

exilis Sanderson, 1938a

Stenelmis exilis Sanderson, 1938a: 680

Type Locality: USA (Arkansas).

Distribution: USA (Arkansas, Missouri, Oklahoma).

fastuosa Delève, 1966a

Stenelmis fastuosa Delève, 1966a: 90

Type Locality: Congo (DR).

Distribution: Congo (DR).

fissicollis Champion, 1923

Stenelmis fissicollis Champion, 1923: 166

Type Locality: India (Uttarakhand).

Distribution: India (Uttarakhand).

formosana Jeng & Yang, 1991

Stenelmis formosana Jeng & Yang, 1991: 242

Type Locality: Taiwan.

Distribution: Taiwan.

fukiensis Bollow, 1941

Stenelmis fukiensis Bollow, 1941: 75

Type Locality: China (Fujian).

Distribution: China (Fujian).

fursovi Zaitzev, 1951

Stenelmis fursovi Zaitzev, 1951: 73

TYPE LOCALITY: Tajikistan.

DISTRIBUTION: Tajikistan.

fuscata Blatchley, 1925

Stenelmis fuscata Blatchley, 1925: 164

TYPE LOCALITY: USA (Florida).

DISTRIBUTION: USA (Arkansas, Florida, Georgia, Illinois, Iowa, Mississippi, Missouri, North Carolina, South Carolina, Texas, Wisconsin).

NOTE: Incorrect original spelling: "*fuscatus*".

gades Hinton, 1941b

Stenelmis gades Hinton, 1941b: 94

TYPE LOCALITY: South Africa.

DISTRIBUTION: South Africa.

gammoni White & Brown, 1976

Stenelmis gammoni White & Brown, 1976: 189

TYPE LOCALITY: USA (North Carolina).

DISTRIBUTION: USA (Georgia, North Carolina, South Carolina, Tennessee, Virginia).

gardneri Hinton, 1941b

Stenelmis gardneri Hinton, 1941b: 84

TYPE LOCALITY: Myanmar.

DISTRIBUTION: Myanmar.

gauglerae Bollow, 1941

Stenelmis gauglerae Bollow, 1941: 28

TYPE LOCALITY: "South China".

DISTRIBUTION: "South China".

NOTE: Incorrect original spelling: "*gaugleri*"; named after a woman, "Sekretärin des Museums G. Frey" (Bollow 1941: 30), and therefore the original spelling must be corrected (ICZN 1999: Art. 31.1.2); incorrect subsequent spelling: "*gauglrei*" (Zhang & Yang 2003: 276).

granulosa Zhang, Yang & Zhang, 2003

Stenelmis granulosa Zhang, Yang & Zhang, 2003: 118

TYPE LOCALITY: China (Guangxi).

DISTRIBUTION: China (Guangxi).

NOTE: Incorrect original spelling: "*granulose*" (lapsus calami, gender ending incorrect) (ICZN 1999: Art. 32.5.1), spelled correctly ("*granulosa*") by Zhang (1994: 21, 41) in his unpublished thesis.

grossa Sanderson, 1938a

Stenelmis grossa Sanderson, 1938a: 686

TYPE LOCALITY: USA (Mississippi).

DISTRIBUTION: Canada (Manitoba, Ontario, Quebec), USA (Alabama, Arkansas, Florida, Illinois, Indiana, Iowa, Kansas, Kentucky, Louisiana, Maine, Maryland, Massachusetts, Michigan, Minnesota, Mississippi, Missouri, Nebraska, New Hampshire, New Jersey, North Carolina, North Dakota, Ohio, Oklahoma, Pennsylvania, South Carolina, Tennessee, Texas, West Virginia, Wisconsin).

grossepunctata Bollow, 1941

Stenelmis grossepunctata Bollow, 1941: 56

TYPE LOCALITY: China (Fujian).

DISTRIBUTION: China (Fujian).

NOTE: Incorrect original spelling: "*grossepunctatus*".

grossimarginata Yang & Zhang, 1995

Stenelmis grossimarginata Yang & Zhang, 1995: 102

TYPE LOCALITY: China (Zhejiang).

DISTRIBUTION: China (Fujian, Zhejiang).

NOTE: Originally described in a key, authorship ascribed to Yang et Zhang (Zhang & Yang 1995: 105, 109); the formal description, which was obviously scheduled for publication in 1995, was published after a delay of two years (L. Ji, pers. comm.) by Zhang, Yang & Li (1997: 229); in the English summary of the formal description by Zhang, Yang & Li (1997: 230) the authorship of this species was reversed ("Zhang et Yang"); Jäch et al. (2006: 438) erroneously ascribed the authorship of *Stenelmis grossimarginata* to Zhang, Yang & Li (1997), and they erroneously listed "*Stenelmis grossimarginata* Yang & Zhang, 1995" as a junior homonym of *Stenelmis grossimarginata* Zhang & Yang, 1995.

grouvellei Alluaud, 1933

Stenelmis grouvellei Alluaud, 1933: 156

TYPE LOCALITY: Gabon.

DISTRIBUTION: Congo (DR), Congo (R), Gabon.

SYNONYM: *Stenelmis bidenticulata* Delève, 1938: 358. — Type locality: Congo (DR). NOTE: Synonymized by Delève (1966a: 82).

guangxinensis Zhang, Yang & Zhang, 2003

Stenelmis guangxinensis Zhang, Yang & Zhang, 2003: 122

TYPE LOCALITY: China (Guangxi).

DISTRIBUTION: China (Guangxi).

gutianshana Zhang & Yang, 1995

Stenelmis gutianshana Zhang & Yang, 1995: 104

TYPE LOCALITY: China (Zhejiang).

DISTRIBUTION: China (Zhejiang).

NOTE: Incorrect subsequent spelling: "*guitianshanensis*"(Zhang & Yang 2003: 279, 281).

hafizi Hinton, 1941b

Stenelmis hafizi Hinton, 1941b: 85

TYPE LOCALITY: India (Meghalaya).

DISTRIBUTION: India (Meghalaya).

hayashii Satô, 1999

Stenelmis hayashii Satô, 1999: 121

TYPE LOCALITY: Japan (Ryukyu Islands).

DISTRIBUTION: Japan (Ryukyu Islands).

hera Hinton, 1941b

Stenelmis hera Hinton, 1941b: 98

TYPE LOCALITY: Cameroon.

DISTRIBUTION: Angola, Cameroon, Congo (DR), Ghana.

heteromorpha Yang & Zhang, 1995

Stenelmis heteromorpha Yang & Zhang, 1995: 106

TYPE LOCALITY: China (Guizhou), according to Zhang (1994: 22, 47).

DISTRIBUTION: China (Guizhou).

NOTE: Originally described very briefly in Zhang & Yang (1995), authorship ascribed to Yang et Zhang (Zhang & Yang 1995: 106, 109); no formal description ever published.

hisamatsui Satô, 1960

Stenelmis hisamatsui Satô, 1960: 253

TYPE LOCALITY: Japan (Ryukyu Islands).

DISTRIBUTION: Guam, Japan (Ryukyu Islands).

huangkengana Yang & Zhang, 1995

Stenelmis huangkengana Yang & Zhang, 1995: 105

Type Locality: China (Fujian).

Distribution: China (Fujian).

Note: Originally described very briefly in Zhang & Yang (1995), authorship ascribed to Yang et Zhang (Zhang & Yang 1995: 105, 109); formal description published by Yang & Zhang (2002: 817); erroneously, the year of publication of this species was cited as "2002" by Jäch et al. (2006: 438).

humerosa Motschulsky, 1860

Stenelmis humerosa Motschulsky, 1860: 50

Type Locality: USA (Alabama, "fleuve Alabama" [Alabama River]) (see Motschulsky 1860: 50); erroneously the type locality was attributed to Tennessee by Sanderson (1938a: 693) (partly) and Brown (1983: 11).

Distribution: USA (Alabama, Delaware, Maryland, Massachusetts, New Jersey, Pennsylvania, South Carolina, Tennessee (Schmude 1992).

Note: Incorrect original spelling: "*humerosus*".

Synonym: *Stenelmis linearis* Zimmermann, 1869: 259. — Type locality: USA (South Carolina), see Sanderson (1938a: 693).

hungerfordi Sanderson, 1938a

Stenelmis hungerfordi Sanderson, 1938a: 690

Type Locality: USA (Florida).

Distribution: USA (Florida).

inaccessoria Yang & Zhang, 2002

Stenelmis inaccessoria Yang & Zhang, 2002: 819

Type Locality: China (Fujian).

Distribution: China (Fujian).

indepressa Yang & Zhang, 1995

Stenelmis indepressa Yang & Zhang, 1995: 103

Type Locality: China (Fujian).

Distribution: China (Fujian, Zhejiang).

Note: Originally described in a key, authorship ascribed to Yang & Zhang (Zhang & Yang 1995: 105); incorrect original spelling: "*indepresa*", changed mandatorily by Yang & Zhang (2002: 821), who provided a formal description; authorship erroneously ascribed to "Zhang & Yang" by Jäch et al. (2006: 438); "*indepressa*" was erroneously listed as a synonym of "*indepresa*" (incorrect original spelling) by Jäch et al. (2006: 438).

indica Grouvelle, 1911b

Stenelmis indica Grouvelle, 1911b: 316

Type Locality: India (Kerala).

Distribution: India (Kerala).

insufficiens Jäch & Kodada, 2006

Stenelmis insufficiens Jäch & Kodada, 2006: 60

Type Locality: China (Zhejiang).

Distribution: China (Zhejiang).

Note: Substitute name for *Stenelmis sinuata* Zhang & Yang, 1995.

Homonym: *Stenelmis sinuata* Zhang & Yang, 1995: 107. Note: Junior primary homonym of *Stenelmis sinuata* Le Conte, 1852.

iranensis Delève, 1970b

Stenelmis iranensis Delève, 1970b: 701

Type Locality: Iran.

Distribution: Iran.

ishiharai Satô, 1964b

Stenelmis ishiharai Satô, 1964b: 31

Type Locality: Japan (Ryukyu Islands).

Distribution: Japan (Ryukyu Islands).

jeanneli Grouvelle, 1920

Stenelmis jeanneli Grouvelle, 1920: 203

Type Locality: Kenya.

Distribution: Congo (DR), Kenya.

kaihuana Zhang & Yang, 1995

Stenelmis kaihuana Zhang & Yang, 1995: 105

Type Locality: China (Zhejiang).

Distribution: China (Chongqing, Zhejiang).

kaszabi Delève, 1968c

Stenelmis kaszabi Delève, 1968c: 159

Type Locality: Vietnam.

Distribution: Vietnam.

klapperichi Bollow, 1941

Stenelmis klapperichi Bollow, 1941: 42

Type Locality: China (Fujian).

Distribution: China (Fujian).

knobeli Sanderson, 1938a

Stenelmis knobeli Sanderson, 1938a: 677

Type Locality: USA (Arkansas).

Distribution: USA (Arkansas, Missouri, Oklahoma, Texas, Wisconsin).

kochi Bollow, 1941

Stenelmis kochi Bollow, 1941: 26

Type Locality: "South China".

Distribution: "South China".

Note: Incorrect subsequent spelling: "*koohi*" (Zhang & Yang 2003: 275).

koreana Satô, 1978

Stenelmis koreana Satô, 1978: 147

Type Locality: Korea (DPR).

Distribution: Korea (DPR), Korea (R), Russia (Far East).

Note: Incorrect original spelling: "*koreanus*".

kuntzeni Bollow, 1941

Stenelmis kuntzeni Bollow, 1941: 71

Type Locality: China (Guangdong).

Distribution: China (Guangdong).

Note: Incorrect subsequent spellings: "*kutzeni*" (Hua 2002: 98), "*kuntreni*" (Zhang & Yang 2003: 275).

laeticollis Delève, 1966a

Stenelmis laeticollis Delève, 1966a: 69

Type Locality: Ivory Coast.

Distribution: Ghana, Ivory Coast, Liberia.

lariversi Schmude, 1999

Stenelmis lariversi Schmude, 1999: 2

Type Locality: USA (Nevada).

Distribution: USA (Nevada).

lateralis Sanderson, 1938a

Stenelmis lateralis Sanderson, 1938a: 672

Type Locality: USA (Arkansas).

Distribution: USA (Arkansas, Kentucky, Louisiana, Mississippi, Missouri, Oklahoma, Tennessee).

leblanci Peyerimhoff, 1929

Stenelmis leblanci Peyerimhoff, 1929: 171

Type Locality: Algeria.

Distribution: Algeria.

liberiana Delève, 1973b

Stenelmis liberiana Delève, 1973b: 305

Type Locality: Liberia.

Distribution: Liberia.

lignicola Schmude & Brown, 1992

Stenelmis lignicola Schmude & Brown, 1992: 583

Type Locality: USA (Alabama).

Distribution: USA (Alabama, Arkansas, Delaware, Florida, Georgia, Illinois, Indiana, Iowa, Kentucky, Maryland, Michigan, Mississippi, Missouri, Nebraska, North Carolina, Oklahoma, Pennsylvania, South Carolina).

lutea Zhang, Su & Yang, 2003a

Stenelmis lutea Zhang, Su & Yang, 2003a: 107

Type Locality: China (Guangxi).

Distribution: China (Fujian, Guangxi, Zhejiang).

magnofoveola Bollow, 1941

Stenelmis magnofoveola Bollow, 1941: 48

Type Locality: China (Fujian).

Distribution: China (Fujian).

Note: Incorrect subsequent spelling: "*magnojoveola*" (Zhang & Yang 2003: 275).

malaisei Bollow, 1940b

Stenelmis malaisei Bollow, 1940b: 22

Type Locality: Myanmar.

Distribution: Myanmar.

mera Sanderson, 1938a

Stenelmis mera Sanderson, 1938a: 682

Type Locality: USA (Tennessee).

Distribution: Canada (New Brunswick, Nova Scotia, Ontario, Quebec), USA (Alabama, Arkansas, Connecticut, Delaware, Georgia, Iowa, Kentucky, Maine, Maryland, Michigan, Missouri, New Hampshire, New York, North Carolina, Oklahoma, Pennsylvania, South Carolina, Tennessee, Virginia, Wisconsin).

merella Hinton, 1941b

Stenelmis merella Hinton, 1941b: 101

Type Locality: Sierra Leone.

Distribution: Sierra Leone.

metatibialis Delève, 1968c

Stenelmis metatibialis Delève, 1968c: 161

Type Locality: Vietnam.

Distribution: Vietnam.

minuta Grouvelle, 1896d

Stenelmis minuta Grouvelle, 1896d: 48

Type Locality: Indonesia (Sumatra).

Distribution: Indonesia (Java, Sumatra).

mirabilis Sanderson, 1938a

Stenelmis mirabilis Sanderson, 1938a: 693

Type Locality: USA (Connecticut).

Distribution: USA (Connecticut).

moapa La Rivers, 1949b

Stenelmis moapa La Rivers, 1949b: 218

Type Locality: USA (Nevada).

Distribution: USA (Nevada).

Note: Epithet name derives from the Piute adjective "moapa" (= muddy), latinized: moapus, -a, -um; originally described as a subspecies of *Stenelmis calida* Chandler, 1949.

montana Yang & Zhang, 1995

Stenelmis montana Yang & Zhang, 1995: 102

Type Locality: China (Guangxi).

Distribution: China (Fujian, Guangxi, Zhejiang).

Note: Originally described in a key by Zhang & Yang (1995), authorship ascribed to "Yang et Zhang" (Zhang & Yang 1995: 106); formal description published by Zhang, Su & Yang (2003a: 108); authorship erroneously ascribed to "Zhang & Yang" by Jäch et al. (2006: 439).

morsei White, 1982

Stenelmis morsei White, 1982: 171

Type Locality: USA (South Carolina).

Distribution: USA (New York, North Carolina, Ohio, Pennsylvania, South Carolina, Virginia, West Virginia).

musgravei Sanderson, 1938a

Stenelmis musgravei Sanderson, 1938a: 698

Type Locality: USA (Virginia).

Distribution: Canada (Ontario, Quebec), USA (Alabama, Florida, Georgia, Illinois, Indiana, Maine, Maryland, Michigan, Mississippi, Missouri, New York, Pennsylvania, Texas, Virginia, West Virginia, Wisconsin).

nematodes Janssens, 1959a

Stenelmis nematodes Janssens, 1959a: 6

Type Locality: Afghanistan.

Distribution: Afghanistan.

nipponica Nomura, 1958a

Stenelmis nipponica Nomura, 1958a: 41

Type Locality: Japan (Shikoku).

Distribution: Japan (Honshu, Kuyshu, Shikoku), Korea (DPR), Korea (R).

noblei Delève, 1966a

Stenelmis noblei Delève, 1966a: 67

Type Locality: South Africa.

Distribution: South Africa.

nomia Hinton, 1941b

Stenelmis nomia Hinton, 1941b: 103

Type Locality: Cameroon.

Distribution: Cameroon.

notabilis Delève, 1968c

Stenelmis notabilis Delève, 1968c: 160

Type Locality: Vietnam.

Distribution: Vietnam.

obscurifusca Zhang, Yang & Zhang, 2003

Stenelmis obscurifusca Zhang, Yang & Zhang, 2003: 119

Type Locality: China (Guangxi).

Distribution: China (Guangxi, Guizhou).

occidentalis Schmude & Brown, 1991

Stenelmis occidentalis Schmude & Brown, 1991: 52

Type Locality: USA (Texas).

Distribution: Mexico, USA (Arkansas, Arizona, Colorado, Louisiana, Montana, Nebraska, Nevada, Oklahoma, Oregon, South Dakota, Texas).

orthotibiata Yang & Zhang, 2002

Stenelmis orthotibiata Yang & Zhang, 2002: 814

Type Locality: China (Fujian).

Distribution: China (Fujian, Guangxi).

Note: Formal description published again by Zhang & Yang (2003: 278).

palawana Delève, 1973c

Stenelmis palawana Delève, 1973c: 29

Type Locality: Philippines (Palawan).

Distribution: Philippines (Palawan).

palembangensis Delève, 1970d

Stenelmis palembangensis Delève, 1970d: 248

Type Locality: Indonesia (Sumatra).

Distribution: Indonesia (Sumatra).

parva Sanderson, 1938a

Stenelmis parva Sanderson, 1938a: 688

Type Locality: USA (Oklahoma).

Distribution: USA (Arkansas, Oklahoma, Texas).

parvifoveola Zhang, Su & Yang, 2003a

Stenelmis parvifoveola Zhang, Su & Yang, 2003a: 107

Type Locality: China (Guangxi).

Distribution: China (Guangxi).

peyerimhoffi Bollow, 1941

Stenelmis peyerimhoffi Bollow, 1941: 15

Type Locality: Morocco.

Distribution: Morocco.

phymatodes Alluaud, 1933

Stenelmis phymatodes Alluaud, 1933: 156

Type Locality: Ivory Coast.

Distribution: Burkina Faso, Congo (DR), Congo (R), Ghana, Guinea, Ivory Coast, Liberia.

planiuscula Delève, 1963f

Stenelmis planiuscula Delève, 1963f: 812

Type Locality: Cameroon.

Distribution: Cameroon, Guinea.

pocsi Delève, 1968c

Stenelmis pocsi Delève, 1968c: 157

Type Locality: Vietnam.

Distribution: Vietnam.

priapus Hinton, 1941b

Stenelmis priapus Hinton, 1941b: 76

Type Locality: China (Fujian).

Distribution: China (Fujian).

Note: Epithet name is a noun in apposition (fertility god in Greek mythology).

prusias Hinton, 1941b

Stenelmis prusias Hinton, 1941b: 96

Type Locality: Malawi, erroneously attributed to South Africa ("Natal") by Delève (1966a: 74).

Distribution: Congo (DR), Guinea, Malawi, Uganda.

puberula Reitter, 1887

Stenelmis puberula Reitter, 1887: 259

Type Locality: Georgia.

Distribution: Afghanistan (Hua 2002: 98), Armenia, Bosnia and Herzegovina, China (Xinjiang)? (Hua 2002: 98), Georgia, Iran, Israel, Lebanon, Russia, Slovakia, Syria, Turkey, Turkmenistan, Ukraine.

Synonyms: *Stenelmis apfelbeki* Kuwert, 1890: 52. — Type locality: Bosnia and Herzegovina. Note: Obviously named for V. Apfelbeck (1859–1934), insect dealer in Sarajevo (Bosnia and Herzegovina), but, besides the name "*apfelbeki*" (appearing in the original publication three times) also the patronym (Apfelbeck) itself was spelled incorrectly ("Apfelbek", without "c") throughout the publication (five times); therefore the name "*apfelbeki*" must be regarded as correct original spelling (ICZN 1999: Art. 32.5.1: "without recourse to any external source of information"); incorrect subsequent spelling: "*apfelbecki*" (Grouvelle 1897: 206, Zaitzev 1908: 300, 1910: 23).

Stenelmis damascena Reitter, 1907: 484. — Type locality: Syria.

punctatissima Bollow, 1940b

Stenelmis punctatissima Bollow, 1940b: 26

Type Locality: Myanmar.

Distribution: Myanmar.

Note: Incorrect original spelling: "*punctatissimus*", herewith changed mandatorily (ICZN 1999: Art. 34.2).

punctulata Bollow, 1940b

Stenelmis punctulata Bollow, 1940b: 24

Type Locality: Myanmar.

Distribution: China (Fujian) (Bollow 1941: 59), Myanmar.

Note: Incorrect original spelling: "*punctulatus*".

quadrimaculata Horn, 1870

Stenelmis quadrimaculata Horn, 1870: 40

Type Locality: USA (Vermont).

Distribution: Canada (Ontario, Quebec), USA (Illinois, Indiana, Maryland, Michigan, Minnesota, New York, North Carolina, Texas, Vermont, Virginia).

Note: Incorrect original spelling: "*quadrimaculatus*"; incorrect subsequent spelling: "*4-maculatus*" (Crotch 1873: 53).

Synonym: *Stenelmis blatchleyi* Musgrave, 1933: 57. Note: Substitute name for *Stenelmis sulcata* Blatchley, 1910.

Homonym: *Stenelmis sulcata* Blatchley, 1910: 681. — Type locality: USA (Indiana). Note: Junior primary homonym of *Stenelmis sulcata* Grouvelle, 1892a (*Leptelmis*); incorrect original spelling: "*sulcatus*".

ritsemae Reitter, 1889

Stenelmis ritsemae Reitter, 1889: 8

Type Locality: Indonesia (Sumatra).

Distribution: Indonesia (Sumatra).

Note: Incorrect subsequent spelling: "*ritsemai*" (Zaitzev 1908: 300, 1910: 23); formed from a personal name (Conrad Ritsema, 1846–1929, Dutch entomologist), which was accepted as a Latin name (see ICZN 1999: Art. 31.1.1).

roiae Bollow, 1941

Stenelmis roiae Bollow, 1941: 60

Type Locality: "South China".

Distribution: "South China".

Note: Incorrect original spelling: "*roii*"; named after a woman, Rosetta Roi (Bollow 1941: 61), and therefore the original spelling must be corrected (ICZN 1999: Art. 31.1.2).

rufocarinata Delève, 1956

Stenelmis rufocarinata Delève, 1956: 374

TYPE LOCALITY: Rwanda.

DISTRIBUTION: Rwanda.

sandersoni Musgrave, 1940

Stenelmis sandersoni Musgrave, 1940: 48

TYPE LOCALITY: USA (Virginia).

DISTRIBUTION: Canada (Ontario, Quebec), USA (Alabama, Arkansas, Connecticut, Georgia, Illinois, Kentucky, Maine, Maryland, Michigan, Minnesota, Missouri, New Jersey, New York, North Carolina, Ohio, Oklahoma, Pennsylvania, South Carolina, Tennessee, Virginia, West Virginia).

NOTE: Substitute name for *Stenelmis tarsalis* Sanderson, 1938a.

HOMONYM: *Stenelmis tarsalis* Sanderson, 1938a: 675. NOTE: Junior primary homonym of *Stenelmis tarsalis* Delève, 1937a.

sauteri Kôno, 1936

Stenelmis sauteri Kôno, 1936: 122

TYPE LOCALITY: Taiwan.

DISTRIBUTION: Taiwan.

scotti Hinton, 1940g

Stenelmis scotti Hinton, 1940g: 304

TYPE LOCALITY: Ethiopia.

DISTRIBUTION: Ethiopia.

scutellicarinata Zhang & Yang, 1995

Stenelmis scutellicarinata Zhang & Yang, 1995: 106

TYPE LOCALITY: China (Zhejiang).

DISTRIBUTION: China (Zhejiang).

semifumosa Hinton, 1936e

Stenelmis semifumosa Hinton, 1936e: 218

TYPE LOCALITY: Malaysia (Sarawak).

DISTRIBUTION: Malaysia (Sarawak).

NOTE: Incorrect original spelling: "*semifumosus*", herewith changed mandatorily (ICZN 1999: Art. 34.2).

semirubra Reitter, 1889

Stenelmis semirubra Reitter, 1889: 8

TYPE LOCALITY: Indonesia (Sumatra).

Distribution: Indonesia (Java, Sumatra).

Note: Incorrect original spelling: "*semirubrum*".

seres Hinton, 1941b

Stenelmis seres Hinton, 1941b: 90

Type Locality: Malaysia (Kedah).

Distribution: Malaysia (Kedah), Vietnam.

Note: Epithet name is a noun in apposition (inhabitants of Serica [silk land]).

sexlineata Sanderson, 1938a

Stenelmis sexlineata Sanderson, 1938a: 663

Type Locality: USA (Kansas).

Distribution: USA (Alabama, Arkansas, Colorado, Illinois, Indiana, Iowa, Kansas, Kentucky, Michigan, Minnesota, Missouri, Nebraska, New Mexico, Ohio, Oklahoma, South Dakota, Tennessee, Texas, Wisconsin).

simplex nyongi Delève, 1963f

Stenelmis simplex nyongi Delève, 1963f: 815

Type Locality: Cameroon.

Distribution: Cameroon.

simplex simplex Delève, 1938

Stenelmis simplex Delève, 1938: 360

Type Locality: Congo (DR).

Distribution: Congo (DR), Congo (R), Liberia.

sinica Yang & Zhang, 1995

Stenelmis sinica Yang & Zhang, 1995: 102

Type Locality: China (Fujian).

Distribution: China (Anhui, Fujian, Gansu, Guizhou, Hubei, Hunan, Sichuan, Zhejiang).

Note: Originally described in a key, authorship ascribed to Yang & Zhang (Zhang & Yang 1995: 107); formal description published by Yang & Zhang (2002: 815); authorship erroneously ascribed to "Zhang & Yang" by Jäch et al. (2006: 439).

sinuata Le Conte, 1852

Stenelmis sinuata Le Conte, 1852: 44

Type Locality: USA (Georgia).

Distribution: USA (Alabama, Delaware, Florida, Georgia, Mississippi, New Jersey, North Carolina, South Carolina, Virginia).

Note: Incorrect original spelling: "*sinuatus*".

soror Delève, 1938

Stenelmis soror Delève, 1938: 362

Type Locality: Congo (DR).

Distribution: Central African Republic, Congo (DR), Ivory Coast, Liberia.

subsulcata Grouvelle, 1896a

Stenelmis subsulcata Grouvelle, 1896a: 44

Type Locality: Indonesia (Sumatra).

Distribution: Indonesia (Sumatra).

subtilis Zaitzev, 1951

Stenelmis subtilis Zaitzev, 1951: 72

Type Locality: Kazakhstan.

Distribution: Kazakhstan.

sulcaticarinata Zhang & Yang, 1995

Stenelmis sulcaticarinata Zhang & Yang, 1995: 102

Type Locality: China (Fujian).

Distribution: China (Fujian, Zhejiang).

Note: Originally described in a key; formal description published by Yang & Zhang (2002: 816).

sulcifrons Bollow, 1940b

Stenelmis sulcifrons Bollow, 1940b: 20

Type Locality: Myanmar.

Distribution: Myanmar.

sulmo Hinton, 1941b

Stenelmis sulmo Hinton, 1941b: 73

Type Locality: China (Fujian).

Distribution: China (Fujian).

Note: Incorrect subsequent spellings: "*salmo*" (Zhang & Yang 1995: 109), "*sulno*" (Yang & Zhang 2002: 815), "*sulma*" (Zhang & Yang 2003: 276).

szechuanensis Mařan, 1939

Stenelmis szechuanensis Mařan, 1939: 40

Type Locality: China (Sichuan).

Distribution: China (Sichuan).

szekessyi Bollow, 1941

Stenelmis szekessyi Bollow, 1941: 46

Type Locality: India (Maharashtra).

Distribution: India (Maharashtra).

tarsalis Delève, 1937a

Stenelmis tarsalis Delève, 1937a: 151

Type Locality: Congo (DR).

Distribution: Congo (DR), Congo (R).

testacea Grouvelle, 1892b

Stenelmis testacea Grouvelle, 1892b: 867

Type Locality: Myanmar.

Distribution: Myanmar, "South China" (Bollow 1941: 21).

thusa Hinton, 1941b

Stenelmis thusa Hinton, 1941b: 99

Type Locality: Uganda.

Distribution: South Africa, Uganda.

trisulcata Fairmaire, 1889a

Stenelmis trisulcata Fairmaire, 1889a: 337

Type Locality: Vietnam.

Distribution: Vietnam, China (Beijing, "South China" (Bollow 1941: 39)).

troilus Hinton, 1941b

Stenelmis troilus Hinton, 1941b: 75

Type Locality: China (Fujian).

Distribution: China (Fujian).

Note: Epithet name is a noun in apposition (Trojan prince).

tros Hinton, 1941b

Stenelmis tros Hinton, 1941b: 77

Type Locality: China (Fujian).

Distribution: China (Fujian).

Note: Epithet name is a noun in apposition (in Greek mythology a ruler of Troy); incorrect subsequent spelling: "*tras*" (Hua 2002: 98, Zhang & Yang 2003: 276).

unguicularis Bollow, 1940b

Stenelmis unguicularis Bollow, 1940b: 17

Type Locality: Myanmar.

Distribution: Myanmar.

venticarinata Zhang & Yang, 1995

Stenelmis venticarinata Zhang & Yang, 1995: 102

Type Locality: China (Guizhou).

Distribution: China (Fujian, Guangxi, Guizhou, Zhejiang).

Note: Originally described in a key; formal description published by Zhang & Yang (2003: 276); incorrect subsequent spelling: "*venticarinate*" (Zhang & Yang 2003: 280).

ventiplana Zhang & Yang, 2003

Stenelmis ventiplana Zhang & Yang, 2003: 276

Type Locality: China (Guangxi).

Distribution: China (Guangxi).

Note: Similar to *Stenelmis nipponica* Nomura, 1958a (Jung et al. 2015: 108).

vicina Grouvelle, 1896d

Stenelmis vicina Grouvelle, 1896d: 47

Type Locality: Indonesia (Engano).

Distribution: Indonesia (Engano).

villosocostata Grouvelle, 1896a

Stenelmis villosocostata Grouvelle, 1896a: 46

Type Locality: Vietnam.

Distribution: Vietnam.

vulgaris Nomura, 1958a

Stenelmis vulgaris Nomura, 1958a: 44

Type Locality: Japan (Honshu).

Distribution: China (Liaoning) (Li 1992: 88), Japan (Honshu, Kyushu, Shikoku), Korea (DPR), Korea (R).

Synonym: *Stenelmis miyamotoi* Nomura & Nakane, 1958: 81. — Type locality: Japan (Honshu). Note: Synonymized by Hayashi et al. (2013: 693).

weyersi Grouvelle, 1900b

Stenelmis weyersi Grouvelle, 1900b: 268

Type Locality: Indonesia (Sumatra).

Distribution: Indonesia (Sumatra).

wongi Jeng & Yang, 1991

Stenelmis wongi Jeng & Yang, 1991: 240

Type Locality: Taiwan.

Distribution: Taiwan.

Note: Incorrect subsequent spelling: "*wonqi*" (Zhang & Yang 2003: 276).

xylonastis Schmude & Barr, 1992

Stenelmis xylonastis Schmude & Barr, 1992: 587

TYPE LOCALITY: USA (Texas).

DISTRIBUTION: USA (Alabama, Arkansas, Delaware, Florida, Georgia, Indiana, Kentucky, Louisiana, Maryland, Mississippi, Missouri, North Carolina, Oklahoma, Pennsylvania, South Carolina, Texas, Virginia).

yangi Zhang & Ding, 1995

Stenelmis yangi Zhang & Ding, 1995: 16

TYPE LOCALITY: China (Beijing).

DISTRIBUTION: China (Beijing).

zhangi Jäch & Kodada, 2006

Stenelmis zhangi Jäch & Kodada, 2006: 60

TYPE LOCALITY: China (Guangxi).

DISTRIBUTION: China (Guangxi).

NOTE: Substitute name for *Stenelmis subsulcata* Zhang, Su & Yang, 2003a: 107; the name of the second author of *Stenelmis zhangi* was incorrectly spelled as "Kodana" in Jäch & Kodada (2006: 60) and Jäch et al. (2006: 439).

HOMONYM: *Stenelmis subsulcata* Zhang, Su & Yang, 2003a: 107. NOTE: Junior primary homonym of *Stenelmis subsulcata* Grouvelle, 1896a; partly misspelled as "*subsuleata*" in original description (p. 106).

Stenhelmoides Grouvelle, 1908

Stenhelmoides Grouvelle, 1908: 182. — Type species: *Stenhelmoides guyanensis* Grouvelle, 1908. — Gender masculine (ICZN 1999: Art. 30.1.4.4).

NOTE: Incorrect subsequent spelling: "*Stenelmoides*" (Zaitzev 1910: 23), not to be regarded as an emendation (ICZN 1999: Art. 33.2.1).

SYNONYM: *Stenelsianus* Hinton, 1934: 196. — Type species: *Stenelsianus rufulus* Hinton, 1934. — Gender masculine.

antennarius Spangler & Perkins, 1989

Stenhelmoides antennarius Spangler & Perkins, 1989: 20

TYPE LOCALITY: Ecuador.

DISTRIBUTION: Ecuador.

beebei Spangler & Perkins, 1989

Stenhelmoides beebei Spangler & Perkins, 1989: 29

TYPE LOCALITY: Guyana.

DISTRIBUTION: Guyana, Venezuela.

grandis Spangler & Perkins, 1989

Stenhelmoides grandis Spangler & Perkins, 1989: 34

Type Locality: Venezuela.

Distribution: Brazil (Amazonas), French Guiana, Venezuela.

grouvellei Pic, 1923

Stenhelmoides grouvellei Pic, 1923: 4 ("*Stenelmoides*")

Type Locality: French Guiana.

Distribution: French Guiana.

guyanensis Grouvelle, 1908

Stenhelmoides guyanensis Grouvelle, 1908: 182

Type Locality: French Guiana.

Distribution: French Guiana, Guyana.

inconscriptus Spangler & Perkins, 1989

Stenhelmoides inconscriptus Spangler & Perkins, 1989: 8

Type Locality: Ecuador.

Distribution: Ecuador.

mimicus Spangler & Perkins, 1989

Stenhelmoides mimicus Spangler & Perkins, 1989: 53

Type Locality: Venezuela.

Distribution: Venezuela.

onorei Monte & Mascagni, 2012

Stenhelmoides onorei Monte & Mascagni, 2012: 23

Type Locality: Ecuador.

Distribution: Ecuador.

platysternum Spangler & Perkins, 1989

Stenhelmoides platysternum Spangler & Perkins, 1989: 39

Type Locality: Ecuador.

Distribution: Ecuador.

Note: Epithet name is a noun in apposition.

pubipes Spangler & Perkins, 1989

Stenhelmoides pubipes Spangler & Perkins, 1989: 17

Type Locality: Ecuador.

Distribution: Ecuador, Peru.

rufulus (Hinton, 1934)

Stenelsianus rufulus Hinton, 1934: 197

Type Locality: Panama.

Distribution: Belize, Colombia, Costa Rica, Ecuador, Guatemala, Honduras, Panama, Peru, Venezuela.

stegastus Spangler & Perkins, 1989

Stenhelmoides stegastus Spangler & Perkins, 1989: 49

Type Locality: Venezuela.

Distribution: Venezuela.

strictifrons Grouvelle, 1908

Stenhelmoides strictifrons Grouvelle, 1908: 183

Type Locality: French Guiana.

Distribution: Brazil (Goiás, Mato Grosso, Pará, São Paulo), French Guiana, Guyana, Peru, Venezuela.

submaculatus Hinton, 1937a

Stenhelmoides submaculatus Hinton, 1937a: 109 ("*Stenelmoides*")

Type Locality: Brazil (Espírito Santo).

Distribution: Brazil (Espírito Santo), Paraguay.

variabilis Spangler & Perkins, 1989

Stenhelmoides variabilis Spangler & Perkins, 1989: 45

Type Locality: Venezuela.

Distribution: Venezuela.

Stethelmis Hinton, 1945d

Stethelmis Hinton, 1945d: 73. — Type species: *Stethelmis chilensis* Hinton, 1945d. — Gender feminine.

chilensis Hinton, 1945d

Stethelmis chilensis Hinton, 1945d: 74

Type Locality: Chile.

Distribution: Chile.

kaszabi Hinton, 1970

Stethelmis kaszabi Hinton, 1970: 110

Type Locality: Argentina.

Distribution: Argentina, Chile.

Stetholus Carter & Zeck, 1929

Stetholus Carter & Zeck, 1929: 52. — Type species: *Stetholus elongatus* Carter & Zeck, 1929. — Gender masculine.

elongatus Carter & Zeck, 1929

Stetholus elongatus Carter & Zeck, 1929: 53

TYPE LOCALITY: Australia (New South Wales).

DISTRIBUTION: Australia (ACT, New South Wales).

Taprobanelmis Delève, 1973a

Taprobanelmis Delève, 1973a: 22. — Type species: *Taprobanelmis carinata* Delève, 1973a. — Gender feminine.

NOTE: Very similar to *Podelmis* Hinton, 1941b.

carinata Delève, 1973a

Taprobanelmis carinata Delève, 1973a: 23

TYPE LOCALITY: Sri Lanka.

DISTRIBUTION: Sri Lanka.

Tolmerelmis Hinton, 1972d

Tolmerelmis Hinton, 1972d: 39. — Type species: *Heterelmis pubipes* Hinton, 1936b. — Gender feminine.

pubipes (Hinton, 1936b)

Heterelmis pubipes Hinton, 1936b: 285

TYPE LOCALITY: Brazil (Santa Catarina).

DISTRIBUTION: Argentina, Brazil (Santa Catarina), Paraguay.

Tolriolus Hinton, 1940a

Tolriolus Hinton, 1940a: 254. — Type species: *Limnius ungulatus* Hinton, 1934. — Gender masculine.

ungulatus (Hinton, 1934)

Limnius ungulatus Hinton, 1934: 200

TYPE LOCALITY: Mexico.

DISTRIBUTION: Mexico.

Trachelminthopsis Delève, 1965a

Trachelminthopsis Delève, 1965a: 110. — Type species: *Trachelminthopsis terrifera* Delève, 1965a. — Gender feminine.

terrifera Delève, 1965a

Trachelminthopsis terrifera Delève, 1965a: 111

Type Locality: Congo (DR).

Distribution: Angola, Congo (DR), Congo (R), Gabon.

Tropidelmis Delève, 1964d

Tropidelmis Delève, 1964d: 537. — Type species: *Tropidelmis hintoni* Delève, 1964d. — Gender feminine.

hintoni Delève, 1964d

Tropidelmis hintoni Delève, 1964d: 538

Type Locality: South Africa.

Distribution: South Africa.

Tyletelmis Hinton, 1972d

Tyletelmis Hinton, 1972d: 37. — Type species: *Tyletelmis mila* Hinton, 1972d. — Gender feminine.

mila Hinton, 1972d

Tyletelmis mila Hinton, 1972d: 37

Type Locality: Brazil (Rondônia); erroneously Hinton (1972d: 39) attributed the type locality to Mato Grosso.

Distribution: Brazil (Amazonas, Pará, Rondônia), French Guiana.

Unguisaeta Jäch, 1982b

Unguisaeta Jäch, 1982b: 94. — Type species: *Unguisaeta rubrica* Jäch, 1982b. — Gender feminine.

rubrica Jäch, 1982b

Unguisaeta rubrica Jäch, 1982b: 95

Type Locality: Sri Lanka.

Distribution: Sri Lanka.

Urumaelmis Satô, 1963a

Urumaelmis Satô, 1963a: 39. — Type species: *Zaitzevia uenoi* Nomura, 1961. — Gender feminine.

NOTE: Originally described as a subgenus of *Zaitzevia* Champion, 1923.

uenoi tokarana (Satô, 1963a)

Zaitzevia uenoi tokarana Satô, 1963a: 40

TYPE LOCALITY: Japan (Ryukyu Islands).

DISTRIBUTION: Japan (Ryukyu Islands).

uenoi uenoi (Nomura, 1961)

Zaitzevia uenoi Nomura, 1961: 2

TYPE LOCALITY: Japan (Ryukyu Islands).

DISTRIBUTION: Japan (Ryukyu Islands).

SYNONYM: *Zaitzevia uenoi amamiana* Satô, 1963a: 41. — Type locality: Japan (Ryukyu Islands).

Vietelmis Delève, 1968c

Vietelmis Delève, 1968c: 174. — Type species: *Vietelmis brevicornis* Delève, 1968c. — Gender feminine.

NOTE: Incorrect subsequent spelling: "*Vietelrnis*" (http://gni.globalnames.org).

brevicornis Delève, 1968c

Vietelmis brevicornis Delève, 1968c: 175

TYPE LOCALITY: Vietnam.

DISTRIBUTION: Laos, Vietnam.

dentipes Kodada & Čiampor, 2000

Vietelmis dentipes Kodada & Čiampor, 2000: 79

TYPE LOCALITY: Malaysia (Sabah).

DISTRIBUTION: Malaysia (Sabah).

ketua Kodada & Čiampor, 2000

Vietelmis ketua Kodada & Čiampor, 2000: 87

TYPE LOCALITY: Malaysia (Sabah).

DISTRIBUTION: Malaysia (Sabah, Sarawak).

kovaci Kodada & Čiampor, 2000

Vietelmis kovaci Kodada & Čiampor, 2000: 67

TYPE LOCALITY: Malaysia (Johor).

DISTRIBUTION: Malaysia (Johor).

lantiri Kodada & Čiampor, 2000

Vietelmis lantiri Kodada & Čiampor, 2000: 78

TYPE LOCALITY: Malaysia (Sabah).

DISTRIBUTION: Malaysia (Sabah).

sinensis Kodada & Čiampor, 2000

Vietelmis sinensis Kodada & Čiampor, 2000: 73

TYPE LOCALITY: China (Hunan).

DISTRIBUTION: China (Guangxi, Hainan, Hunan).

Xenelmis Hinton, 1936c

Xenelmis Hinton, 1936c: 427. — Type species: *Elmis bufo* Sharp, 1882. — Gender feminine.

audax Hinton, 1946a

Xenelmis audax Hinton, 1946a: 239

TYPE LOCALITY: Argentina.

DISTRIBUTION: Argentina, Brazil (Santa Catarina).

bufo (Sharp, 1882)

Elmis bufo Sharp, 1882: 140

TYPE LOCALITY: Panama.

DISTRIBUTION: Belize, Costa Rica, Guatemala, Mexico, Panama, Venezuela.

comis Hinton, 1946a

Xenelmis comis Hinton, 1946a: 240

TYPE LOCALITY: Paraguay.

DISTRIBUTION: Argentina, Brazil (Santa Catarina), Paraguay.

granata (Grouvelle, 1889b)

Helmis granata Grouvelle, 1889b: 405

TYPE LOCALITY: Brazil (Santa Catarina).

DISTRIBUTION: Brazil (Amazonas, Rio de Janeiro, Santa Catarina).

laura Brown, 1970b

Xenelmis laura Brown, 1970b: 62

Type Locality: Brazil (Pará).

Distribution: Brazil (Pará).

leechi Perkins & Steiner, 1981

Xenelmis leechi Perkins & Steiner, 1981: 306

Type Locality: Peru.

Distribution: Bolivia (Manzo 2013: 212), Peru.

marcapata Perkins & Steiner, 1981

Xenelmis marcapata Perkins & Steiner, 1981: 309

Type Locality: Peru.

Distribution: Peru.

micros (Grouvelle, 1889b)

Helmis micros Grouvelle, 1889b: 406

Type Locality: Brazil (Santa Catarina).

Distribution: Brazil (Santa Catarina), Paraguay.

rufipes Delève, 1968a

Xenelmis rufipes Delève, 1968a: 233

Type Locality: Ecuador.

Distribution: Colombia, Ecuador.

sandersoni Brown, 1985

Xenelmis sandersoni Brown, 1985: 53

Type Locality: USA (Arizona).

Distribution: Mexico, USA (Arizona).

tarsalis Hinton, 1940a

Xenelmis tarsalis Hinton, 1940a: 297

Type Locality: Brazil (Rondônia); erroneously, Hinton (1946a: 238) attributed the type locality to Mato Grosso ("Matto [sic] Grosso, Porto Velho, viii-ix.37").

Distribution: Argentina, Brazil (Rondônia); erroneously reported from Mato Grosso by Segura et al. (2013: 44); in Brazil, this species is known only from the type locality.

Note: Originally described in a key; formal description published by Hinton (1946a: 237).

teres Hinton, 1946a

Xenelmis teres Hinton, 1946a: 238

Type Locality: Brazil (Rondônia); erroneously, Hinton (1946a: 239) attributed the type locality to Mato Grosso ("Matto [sic] Grosso, Porto Velho, viii-ix.37").

DISTRIBUTION: Brazil (Rondônia); erroneously reported from Mato Grosso by Segura et al. (2013: 44); this species is known only from the type locality.

uruzuensis Manzo, 2006

Xenelmis uruzuensis Manzo, 2006: 55

TYPE LOCALITY: Argentina.

DISTRIBUTION: Argentina, Brazil (Goiás), Paraguay.

Xenelmoides Hinton, 1936a

Xenelmoides Hinton, 1936a: 5. — Type species: *Elmis simplex* Darlington, 1927. — Gender masculine (ICZN 1999: Art. 30.1.4.4).

simplex (Darlington, 1927)

Helmis simplex Darlington, 1927: 96

TYPE LOCALITY: Cuba.

DISTRIBUTION: Cuba.

Zaitzevia Champion, 1923

Zaitzevia Champion, 1923: 170. — Type species: *Zaitzevia solidicornis* Champion, 1923. — Gender feminine.

SYNONYM: *Awadoronus* Kôno, 1934: 127. — Type species: *Awadoronus awanus* Kôno, 1934. — Gender masculine.

SUBGENUS: ***Suzevia*** Brown, 2001: 203. — Type species: *Zaitzevia posthonia* Brown, 2001. — Gender feminine.

aritai Satô, 1963a

Zaitzevia aritai Satô, 1963a: 40

TYPE LOCALITY: Japan (Ryukyu Islands).

DISTRIBUTION: Japan (Ryukyu Islands).

NOTE: Originally described in *Zaitzevia* subgenus *Urumaelmis*.

awana (Kôno, 1934)

Awadoronus awanus Kôno, 1934: 128

TYPE LOCALITY: Japan (Shikoku).

DISTRIBUTION: Japan (Honshu, Kyushu, Shikoku), Korea (R).

babai Nomura, 1963

Zaitzevia babai Nomura, 1963: 43

Type Locality: Taiwan.

Distribution: Taiwan.

bhutanica Satô, 1977b

Zaitzevia bhutanica Satô, 1977b: 202

Type Locality: Bhutan.

Distribution: Bhutan.

elongata Nomura, 1962

Zaitzevia elongata Nomura, 1962: 49

Type Locality: Japan (Ryukyu Islands).

Distribution: Japan (Ryukyu Islands).

formosana Nomura, 1963

Zaitzevia formosana Nomura, 1963: 50

Type Locality: Taiwan.

Distribution: Taiwan.

Note: In the original description, this species is described as being "very closely allied to *Zaitzevia acutangula* Champion", which now belongs to *Indosolus* Bollow, 1940b.

malaisei Bollow, 1940b

Zaitzevia malaisei Bollow, 1940b: 29

Type Locality: Myanmar.

Distribution: Myanmar.

nitida Nomura, 1963

Zaitzevia nitida Nomura, 1963: 46

Type Locality: Japan (Honshu).

Distribution: Japan (Honshu).

parallela Nomura, 1963

Zaitzevia parallela Nomura, 1963: 45

Type Locality: Taiwan.

Distribution: Taiwan.

parvula parvula (Horn, 1870)

Macronychus parvulus Horn, 1870: 41

Type Locality: USA (California).

Distribution: Canada (Alberta, British Columbia, Yukon Territory), Mexico, USA (Arizona, California, Colorado, Idaho, Montana, Nevada, New Mexico, Oregon, South Dakota, Utah, Washington, Wyoming).

NOTE: Incorrect subsequent spelling: "*Zaitzevia parvulus*" (Sanderson 1938b: 146, Poole & Gentili 1996: 265).

SYNONYM: *Elmis columbiensis* Angell, 1892: 84. — Type locality: Canada (British Columbia). NOTE: Synonymized by Sanderson (1938b: 146).

parvula thermae (Hatch, 1938)

Macronychus thermae Hatch, 1938: 18

TYPE LOCALITY: USA (Montana).

DISTRIBUTION: USA (Montana).

NOTE: Incorrect subsequent spelling: "*termae*" (Poole & Gentili 1996: 265); originally described as discrete species, downgraded to subspecies by Brown (2001: 203); raised to species level by Poole & Gentili (1996: 265) without presenting any evidence; might be a valid species or a synonym (hot spring variety) of *Zaitzevia parvula*, but probably not a subspecies.

pocsi Delève, 1968c

Zaitzevia pocsi Delève, 1968c: 177

TYPE LOCALITY: Vietnam.

DISTRIBUTION: Vietnam.

posthonia Brown, 2001

Zaitzevia posthonia Brown, 2001: 205 (subgenus *Suzevia*)

TYPE LOCALITY: USA (Oregon).

DISTRIBUTION: Canada (British Columbia), USA (California, Idaho, Oregon, Washington).

rivalis Nomura, 1963

Zaitzevia rivalis Nomura, 1963: 48

TYPE LOCALITY: Japan (Honshu).

DISTRIBUTION: Japan (Honshu, Kyushu).

NOTE: Incorrect subsequent spelling: "*revalis*" (Satô 1977c: 6).

rufa Nomura & Baba, 1961

Zaitzevia rufa Nomura & Baba, 1961: 5

TYPE LOCALITY: Japan (Honshu).

DISTRIBUTION: Japan (Honshu).

solidicornis Champion, 1923

Zaitzevia solidicornis Champion, 1923: 170

TYPE LOCALITY: India (Uttarakhand).

DISTRIBUTION: India (Uttarakhand).

tsushimana Nomura, 1963

Zaitzevia tsushimana Nomura, 1963: 47

TYPE LOCALITY: Japan (Tsushima).

DISTRIBUTION: China (Jilin), Japan (Tsushima), Korea (R) (incl. Jeju Island), Russia (Far East).

yaeyamana Satô, 1963a

Zaitzevia yaeyamana Satô, 1963a: 40

TYPE LOCALITY: Japan (Ryukyu Islands).

DISTRIBUTION: Japan (Ryukyu Islands).

NOTE: Originally described in *Zaitzevia* subgenus *Urumaelmis*.

Zaitzeviaria Nomura, 1959

Zaitzeviaria Nomura, 1959: 35. — Type species: *Zaitzevia brevis* Nomura, 1958b. — Gender feminine.

NOTE: Incorrect subsequent spelling: "*Zeitzeviaria*" (Hayashi et al. 2013: 693).

atratula (Grouvelle, 1911a)

Microdes atratulus Grouvelle, 1911a: 252

TYPE LOCALITY: China (Yunnan).

DISTRIBUTION: China (Yunnan).

bicolor (Pic, 1923)

Grouvelleus bicolor Pic, 1923: 4

TYPE LOCALITY: Vietnam.

DISTRIBUTION: Vietnam.

SYNONYM: *Zaitzeviaria fusca* Delève, 1968c: 178. — Type locality: Vietnam. NOTE: Synonymized by Delève (1970d: 270).

brevis (Nomura, 1958b)

Zaitzevia brevis Nomura, 1958b: 49

TYPE LOCALITY: Japan (Shikoku).

DISTRIBUTION: Japan (Hokkaido, Honshu, Kyushu, Shikoku), Russia (Kunashir) (Palatov 2014: 517, as *Zaitzeviaria gotoi* (Nomura, 1959)).

elongata Jäch, 1982b

Zaitzeviaria elongata Jäch, 1982b: 103

TYPE LOCALITY: Sri Lanka.

DISTRIBUTION: Sri Lanka.

gotoi (Nomura, 1959)

Zaitzevia gotoi Nomura, 1959: 36

Type Locality: Japan (Honshu).

Distribution: Japan (Honshu, Kyushu, Shikoku); the record from Russia (Kunashir) by Palatov (2014: 517) probably refers to *Zaitzeviaria brevis* (Nomura, 1958b).

kuriharai Kamite, Ogata & Satô, 2006

Zaitzeviaria kuriharai Kamite, Ogata & Satô, 2006: 149

Type Locality: Japan (Tsushima).

Distribution: Japan (Tsushima).

laevicollis Delève, 1968c

Zaitzeviaria laevicollis Delève, 1968c: 179

Type Locality: Vietnam.

Distribution: Vietnam.

minuscula (Grouvelle, 1892a)

Macronychus minusculus Grouvelle, 1892a: 187

Type Locality: Indonesia (Sumatra).

Distribution: Indonesia (Sumatra).

minuta Satô, 1977c

Zaitzeviaria minuta Satô, 1977c: 191

Type Locality: Malaysia (Sarawak).

Distribution: Malaysia (Sarawak).

negros Satô, 1977c

Zaitzeviaria negros Satô, 1977c: 193

Type Locality: Philippines (Negros).

Distribution: Philippines (Negros).

ovata (Nomura, 1959)

Zaitzevia ovata Nomura, 1959: 35

Type Locality: Japan (Honshu).

Distribution: Japan (Honshu, Shikoku).

pilosella Delève, 1968c

Zaitzeviaria pilosella Delève, 1968c: 179

Type Locality: Vietnam.

Distribution: Vietnam.

zeylanica Jäch, 1982b

Zaitzeviaria zeylanica Jäch, 1982b: 104

Type Locality: Sri Lanka.

Distribution: Sri Lanka.

Species incertae sedis

The five species listed below were described from South America (Bolivia, Brazil, Colombia) by A. Grouvelle (1843–1917) and V. Motschoulsky (1810–1871), who placed them in the genus *Elmis* (resp. *Helmis*, junior synonym of *Elmis*). However, the genus *Elmis* is confined to the western Palearctic Realm. The original descriptions do not include unambiguous generic characters, and therefore these species cannot be identified and they cannot be properly assigned to any South American genus. The type material was not retrieved so far. According to A. Mantilleri (email of 11.II.2014) the types of the three species described by Grouvelle are not deposited in the Muséum national d'Histoire naturelle, Paris, France. The types of the two species described by Motschoulsky are not deposited in the Zoological Museum of Moscow University, Russia (A. Prokin, email of 13.IV.2015) and they are not found in the Zoological Museum of the Zoological Institute of the Russian Academy of Sciences, St. Petersburg, Russia (A. Kovalev, email of 16.IV.2015). If the type specimens cannot be retrieved, these five names, although being currently available, should be removed from the Official List of Specific Names in Zoology.

Elmis aequinoxialis Motschoulsky, 1851

Elmis aequinoxialis Motschoulsky, 1851: 655

Type Locality: Colombia.

Distribution: Colombia.

Note: Incorrect original spelling: "*æquinoxialis*"; this name was not listed in any publication/catalogue since Motschoulsky (1869: 32).

Elmis cervina (Grouvelle, 1896a)

Helmis cervina Grouvelle, 1896a: 51

Type Locality: Bolivia.

Distribution: Bolivia.

Elmis emiliae (Grouvelle, 1889b)

Helmis emiliae Grouvelle, 1889b: 401

Type Locality: Brazil (Rio de Janeiro).

Distribution: Brazil (Rio de Janeiro).

Note: Incorrect original spelling: "*emiliæ*".

Elmis longior (Grouvelle, 1896a)

Helmis longior Grouvelle, 1896a: 51

TYPE LOCALITY: Bolivia.

DISTRIBUTION: Bolivia.

NOTE: This species might belong to *Heterelmis* Sharp, 1882 (Manzo & Moya 2010: 136).

Elmis ovatula Motschoulsky, 1851

Elmis ovatula Motschoulsky, 1851: 655

TYPE LOCALITY: Colombia.

DISTRIBUTION: Colombia.

NOTE: This name was not listed in any publication/catalogue since Motschoulsky (1869: 32).

Fossil taxa

Five fossil species are currently placed in Elmidae. However, the descriptions of two species (*Limnius perrini* Theobald, 1937, *Potamophilites angustifrons* Haupt, 1956) are based on fragments, which provide no morphological evidence for their placement in Byrrhoidea/Dryopoidea. They should be removed from the Official List of Generic/Specific Names in Zoology.

The elmid larva described by Hayashi & Okushima (2012) from the Miocene might belong to *Stenelmis* Dufour, 1835.

Genera:

Palaeoriohelmis Bollow, 1940a

Palaeoriohelmis Bollow, 1940a: 117. — Type species: *Palaeoriohelmis samlandica* Bollow, 1940a (= *Heterlimnius samlandicus* (Bollow, 1940a)). — Gender feminine.

NOTE: Originally described in Dryopidae ("Helminide [sic] (Col. Dryop.)"); junior synonym of *Heterlimnius* Hinton, 1935 (Bukejs et al. 2015: 453).

Potamophilites Haupt, 1956

Potamophilites Haupt, 1956: 49. — Type species: *Potamophilites angustifrons* Haupt, 1956. — Gender masculine (ICZN 1999: Art. 30.1.4.4).

NOTE: Originally described in Dryopidae; described as being close to *Potamophilus* Germar, 1811 and therefore should be transferred to Elmidae; but characters shown in original description do not provide clear evidence for placement in Byrrhoidea/Dryopoidea.

Species:

Elmis decoratus (Statz, 1939)

Helmis decoratus Statz, 1939: 68

TYPE LOCALITY: Germany / Oligocene.

Heterelmis groehni Bukejs, Alekseev & Jäch, 2015

Heterelmis groehni Bukejs, Alekseev & Jäch, 2015: 456

Type Locality: Russia / Eocene[1].

Heterlimnius samlandicus (Bollow, 1940a)

Palaeoriohelmis samlandica Bollow, 1940a: 118

Type Locality: Russia / Eocene[1].

Note: Originally described in Dryopidae ("Helminide [sic] (Col. Dryop.)"); transferred to *Heterlimnius* by Bukejs et al. (2015: 543).

Limnius perrini Theobald, 1937

Limnius perrini Theobald, 1937: 284

Type Locality: France / Oligocene.

Note: Placement in Byrrhoidea/Dryopoidea most doubtful.

Potamophilites angustifrons Haupt, 1956

Potamophilites angustifrons Haupt, 1956: 49

Type Locality: Germany / Eocene.

Note: Originally described in Dryopidae; described as being close to *Potamophilus* Germar, 1811 and therefore should be transferred to Elmidae; but characters shown in original description do not provide clear evidence for placement in Byrrhoidea/Dryopoidea.

Taxa removed from Elmidae

Extant Taxa:

Betelmis Matsumura, 1915: 214

This genus was described in Parnidae [Dryopidae], but in the formal description provided by Matsumura (1916) it is characterized as being "Near *Stenelmis* Duf." (= Elmidae, which were usually included in Dryopidae at that time); synonym of *Mataeopsephus* Waterhouse, 1876 [Psephenidae].

Betelmis japonica Matsumura, 1915: 214

This species belongs to *Mataeopsephus* Waterhouse, 1876 [Psephenidae].

Erichia Reitter, 1895: 79

This genus was described in "Parnidae: Larini" [Elmidae: Larainae]; transferred to Limnichidae by Jäch & Pütz (2001).

1 For dating of Baltic amber, see Aleksandrova & Zaporozhets (2008).

Erichia longicornis Reitter, 1895: 80

This species was described in "Parnidae: Larini" [Elmidae: Larainae]; transferred to Limnichidae by Jäch & Pütz (2001).

Fossil Taxa:

Elmadulescens Peris, Maier & Sánchez-García, 2015: 283

This genus was removed from Elmidae by Bukejs et al. (2015: 459).

Elmadulescens rugosus Peris, Maier & Sánchez-García, 2015: 283

This species was removed from Elmidae by Bukejs et al. (2015: 459).

Unavailable names

a) Nomina nuda (unavailable according to ICZN 1999: Art. 12, Art. 13, or Art. 16.1).

Most of the nomina nuda were published in the 19th century, mainly in catalogues (e.g. Dejean 1821, 1833, 1836, Sturm 1826, 1843). Some of these names were made available later on. Additional nomina nuda may be discovered in various old catalogues in the future.

Genera:

Ceralimnius: first published by Hinton (1939a: 143); it refers to *Heterlimnius* Hinton, 1935 (see also Sanderson 1954: 2).

Praelimnia (†): published by Theobald (1937: 284); authorship ascribed to Perrin; it refers to *Limnius* Illiger, 1802.

Species:

Elmis acuminata: published by Motschulsky (1853: 13).

Elmis areolata: published by Motschulsky (1853: 13); authorship ascribed to Mannerheim.

Elmis bivittata: published by Dejean (1833: 131, 1836: 145), incorrectly spelled as "*bivittatus*"; it might refer to *Dubiraphia bivittata* (Le Conte, 1852).

Elmis chrysomela: published by Bertolini (1874: 102); authorship ascribed to Marsham.

Elmis confluens: published by Motschulsky (1853: 13); authorship ascribed to Müller.

Elmis dilatata: published by Motschulsky (1853: 13); authorship ascribed to Mannerheim.

Elmis (*Lareynia*) *grouvellei*: published by Reitter (1901: 59); it refers to *Elmis perezi* Heyden, 1870.

Elmis lithophila: published by Sturm (1826: 138), incorrectly spelled as "*lithophilus*"; authorship ascribed to Melsheimer; the epithet name has originally been published as nomen nudum in the genus *Parnus* (Dryopidae) by Melsheimer (1806: 16); it refers to *Helichus lithophilus* Germar, 1824 (Dryopidae).

Elmis mannerheimii: published by Dejean (1836: 146).

Elmis muelleri: published by Motschulsky (1853: 13); authorship ascribed to Mannerheim.

Elmis nigritus: published by Mulsant & Rey (1872: 42); authorship ascribed to Chevrolat; it refers to *Dupophilus brevis* Mulsant & Rey, 1872.

Elmis nitida: published by Dejean (1833: 131, 1836: 146), incorrectly spelled as "*nitidus*"; it might refer to *Riolus nitens* (Müller, 1817).

Elmis rufipes: published by Dejean (1833: 131, 1836: 146).

Elmis scabricollis: published by Dejean (1821: 49).

Elmis tibialis: published by Zaitzev (1947: 90); it refers to *Limnius opacus* Müller, 1806a.

Esolus herthae: first published by Janssens (1955: 2) as „*Esolus Herthæ*"; authorship ascribed to Bollow by Janssens (1955: 2) and to Motschulsky by Mascagni & Calamandrei (1992: 130); it probably refers to *Esolus angustatus* (Müller, 1821) (Mascagni & Calamandrei 1992: 130).

Esolus lantosquensis: first published by Pic (1898: 154), authorship ascribed to Grouvelle; Bertolini (1900: 26) listed it as good species ("*lantosquensis* Gruv. [sic]"), Pic (1900b: 60) listed it as a synonym of *Esolus galloprovincialis* Abeille de Perrin, 1900, Ganglbauer (1904a: 115), Heyden et al. (1906: 373), Caillol (1913: 303), Olmi (1969: 130–131), Berthélemy (1979: 37), and Mascagni & Calamandrei (1992: 130) listed it under *Esolus angustatus* (Müller, 1821).

Heterelmis transversa: published by Grouvelle (1889c: 164).

Limnius cothurnatus: first published by Müller (1806a: 192) and later under the name *Elophorus cothurnatus* by Müller (1806b: 215); authorship ascribed to Zenker by Gemminger & Harold (1868: 937); this name refers to *Macronychus quadrituberculatus* Müller, 1806b.

Limnius nigripes: published by Motschulsky (1853: 13).

Limnius sinaiticus: published by Crotch (1872) and various other authors (see Berthélemy 1979: 74); it refers to *Oulimnius aegyptiacus* (Kuwert, 1890).

Limnius tibialis: published by Motschulsky (1853: 13).

Macronychus armeniacus: published by Motschulsky (1853: 13) and Zaitzev (1947: 90); it refers to *Grouvellinus caucasicus* Victor, 1839.

Macronychus bituberculatus: published by Villa & Villa (1835: 41), and under the name *Elmis bituberculata* (incorrect spelling: "*bituberculatus*") by Dejean (1833: 131, 1836: 146), Erichson (1847: 535), Sturm (1857: 34) and Bertolini (1874: 102); authorship ascribed to Bonelli; it refers to *Stenelmis canaliculata* Gyllenhal, 1808.

Macronychus longimanus: published by Sturm (1826: 167); authorship ascribed to Melsheimer; the epithet name has originally been published as a nomen nudum in the genus *Parnus* (Dryopidae) by Melsheimer (1806: 16); it obviously refers to *Macronychus glabratus* Say, 1825.

Macronychus vittatus: published by Sturm (1843: 99); the epithet name has originally been published as a nomen nudum in the genus *Parnus* (Dryopidae) by Melsheimer (1806: 16); it might refer to *Dubiraphia vittata* (Melsheimer, 1844).

Praelimnia longipennis (†): published by Theobald (1937: 284); authorship ascribed to Perrin; it refers to *Limnius perrini* Theobald, 1937 (†).

Stenelmis hurvatana: published by Zaitzev (1951: 73); authorship ascribed to Fursov; it refers to *Stenelmis fursovi* Zaitzev, 1951.

Stenelmis mexicana: published by Stewart, Friday & Rhame (1973: 960); it refers to *Stenelmis occidentalis* Schmude & Brown, 1991.

Stenelmis taiwana: published by Zhang & Yang (2003: 276, 278), partly misspelled as "*tanwana*" (p. 276), authorship ascribed to Jeng & Yang; obviously this names refers to *Stenelmis formosana* Jeng & Yang, 1991.

Stenelmis variabilis: published by Motschulsky (1853: 13); authorship ascribed to Leach.

b) Infrasubspecific name (unavailable according to ICZN 1999: Art. 45.5–6). As for the determination of subspecific or infrasubspecific rank refer to Lingafelter & Nearns (2013).

Elmis maugetii var. *hungarica* Knie, 1978: 62.

c) Names used in theses (unavailable according to ICZN 1999: Art. 8.1.1)
In the last decades, several taxonomic theses on Elmidae have been published (e.g. Čiampor 2002b, Fernandes 2010, Jäch 1982a, Plachý 2006, Schmude 1992, Zhang 1994). New taxa "described" in theses are not explicitely excluded from nomenclature according to the ICZN (1999). Theses are usually printed on paper and often distributed to various colleagues and museums. The thesis of Fernandes (2010) can even be found in the internet (http://insetosaquaticos.inpa.gov.br/index.php/publicacoes/dissertacoes/2010-1/95-taxonomia-de-elmidae-insecta-coleoptera-no-municipio-de-presidente-figueiredo-amazonas/file). However, Art. 8.1.1 states, that a published work "must be issued for the purpose of providing a public and permanent scientific record". Since many of the taxa "described" in the theses listed above were made available formally in various journals later on (under the same or a similar name, e.g. *Stegoelmis figueirensis* [Fernandes 2010] was published under the name *Stegoelmis figueiredoensis* by Fernandes, Passos & Hamada 2011: 58), it must be assumed that the names published in these theses were not (!) issued for the purpose of providing a public and permanent scientific record, and therefore they should be regarded as unavailable for nomenclatorial purposes.

All names of undescribed species used by Jäch (1982a) were made available simultaneously by Jäch (1982b).

In the taxonomic thesis on the genus *Leptelmis* by Plachý (2006) no species names, only numbers, were used for undescribed species.

The list below includes only those names, which have been described later under a different name, or which have so far not been made available.

Graphelmis huliki Čiampor, 2002b. — Type locality: Philippines (= *G. elisabethjaechae* Čiampor, 2005a).

Graphelmis suteki Čiampor, 2002b. — Type locality: Philippines (= *G. schoedli* Čiampor, 2005a).

Heterelmis browni Fernandes, 2010. — Type locality: Brazil (Amazonas).

Hexacylloepus hintoni Fernandes, 2010. — Type locality: Brazil (Amazonas).

Leptelmis guangxiana Zhang, 1994. — Type locality: China.

Leptelmis incarinata Zhang, 1994. — Type locality: China.

Leptelmis jujana Zhang, 1994. — Type locality: China.

Leptelmis taiwansis Zhang, 1994. — Type locality: Taiwan (proposed as substitue name for *Leptelmis formosana* Nomura, 1962).

Leptelmis vittata qiana/qinana Zhang, 1994. — Type locality: China.

Leptelmis yunnana Zhang, 1994. — Type locality: China.

Neoelmis variecarinata Fernandes, 2010. — Type locality: Brazil (Amazonas).

Pilielmis figueirensis Fernandes, 2010. — Type locality: Brazil (Amazonas).

Pseudamophilus luxianus Zhang, 1994. — Type locality: China.

Stenelmis acricarinata Zhang, 1994. — Type locality: China.

Stenelmis angustihumera Zhang, 1994. — Type locality: China.

Stenelmis chengduana Zhang, 1994. — Type locality: China.

Stenelmis chuangana/chuongana Zhang, 1994. — Type locality: China.

Stenelmis dura Zhang, 1994. — Type locality: China.

Stenelmis florala Schmude, 1992. — Type locality: USA (Alabama).

Stenelmis guilinana Zhang, 1994. — Type locality: China.

Stenelmis gutianensis Zhang, 1994. — Type locality: China.

Stenelmis harleyi Schmude, 1992. — Type locality: USA (South Carolina).

Stenelmis interrupta Zhang, 1994. — Type locality: China.

Stenelmis jinghongana Zhang, 1994. — Type locality: China.

Stenelmis longicarinata Zhang, 1994. — Type locality: China.

Stenelmis longipenis Zhang, 1994. — Type locality: China.

Stenelmis scutelligranata Zhang, 1994. — Type locality: China.

Stenelmis serracaniata Zhang, 1994. — Type locality: China.

Stenelmis spangleri Schmude, 1992. — Type locality: Mexico.

Stenelmis transversisulcata Zhang, 1994. — Type locality: China.

Stenelmis weiweii Zhang, 1994. — Type locality: China.

Stenelmis williami Schmude, 1992. — Type locality: USA (Maine).

Stenelmis yanshana Zhang, 1994. — Type locality: China.

Stenhelmoides longifibulus Fernandes, 2010. — Type locality: Brazil (Amazonas).

Stenhelmoides pilitarsus Fernandes, 2010. — Type locality: Brazil (Amazonas).

Stenhelmoides spangleri Fernandes, 2010. — Type locality: Brazil (Amazonas).

Stenhelmoides spinipenis Fernandes, 2010. — Type locality: Brazil (Amazonas).

Nomina exclusa (taxa intentionally excluded from this catalogue)

In 2005, 2007, 2008 and 2011 Mr. "D.M." published four articles, describing three new genera and four new species of Elmidae from Suriname (South America). A fifth paper, dated 2006 (occasionally cited in D.M.'s publications), probably does in fact not exist.

The descriptions of these seven taxa are very short, not providing characters to enable their recognition. Very probably, all the three genera described by Mr. "D.M." represent in fact well known genera.

The types of the species described by Mr. "D.M." from Suriname are not accessible (depository not even mentioned in one case).

One of the papers (2011) was "coauthored" by Miss Somayeh Ezzatpanah (Tehran, Iran). In fact Miss Ezzatpanah had not been aware that "D.M." had added her as a coauthor, and eventually she informed us that "D.M." had been stalking her for quite a long time, after he saw her photograph in the internet. Unfortunately, Miss Ezzatpanah later faced problems with her university due to the (entirely unintended) cooperation with Mr. "D.M." (at least one of these "joint" articles contained very insulting and vilifying sentences). It must be assumed that all papers published under the name Ezzatpanah (as single author) were in fact solely written and submitted by "D.M.".

Until today "D.M." published about 150 taxonomic papers (about 90% in one, unreviewed Australian journal) on 45 animal families:

Heteroptera: Lygaeidae, Schizopteridae

Coleoptera: Anobiidae, Bothrideridae, Carabidae, Chrysomelidae, Ciidae, Cleridae, Curculionidae, Dryopidae, Dytiscidae, Elateridae, Elmidae, Georissidae, Geotrupidae, Haliplidae, Hydraenidae, Hydrochidae, Hydrophilidae, Lampyridae, Latridiidae, Lymexylidae, Malachiidae, Meloidae, Scarabaeidae, Scirtidae, Scydmaenidae, Staphylinidae, Tenebrionidae, Trogossitidae

Hymenoptera: Formicidae, Oonopidae, Scoliidae

Strepsiptera: Myrmecolacidae

Diptera: Syrphidae

Lepidoptera: Pterophoridae

Acari: Hydrachnidia (family names not mentioned by the author): Arrenuridae, Hydryphantidae, Unionicolidae

Araneae: Caponiidae, Salticidae, "Hawkeswoodidae" (name unavailable)

Opiliones (family not mentioned)

Diplopoda: Glomeridae, Haplodesmidae

Gastropoda: Bulimulidae

Numerous papers have been written about the unacceptable quality of these papers and their destabilizing effects for taxonomy and nomenclature: e.g.

Hansen (1999: 51ff.), Jäch (2000, 2006, 2007), Jäch & Short (2009), Vondel (2003). See also:

http://species.wikimedia.org/wiki/Dewanand_Makhan

http://scholarlyoa.com/2013/04/23/taxonomy/

http://www.antwiki.org/wiki/Makhan,_Dewanand

http://scienceblogs.com/myrmecos/2007/12/13/the-rogue-taxonomist/

Therefore we decided to exclude these taxa from our catalogue. We will not list their names and we will not take any further taxonomic actions concerning these names.

For the same reasons, the taxa described by Mr. "D.M." have not been listed in the Catalogue of the Elmidae of the Neotropical Region (Segura et al. 2013) or in any other synoptic contributions to the Neotropical elmid fauna (e.g. Manzo 2013).

Sondermann (2012) created the term "nomina seminuda" for taxa described by Mr. "D.M.".

Protelmidae

Four genera and six species are currently recognized. Deadline: 31.XII.2014.

Caenelmis Spangler, 1996a

Caenelmis Spangler, 1996a: 19. — Type species: *Caenelmis octomeria* Spangler, 1996a. — Gender feminine.

octomeria Spangler, 1996a

Caenelmis octomeria Spangler, 1996a: 19

Type Locality: Kenya.

Distribution: Kenya.

Haplelmis Delève, 1964b

Haplelmis Delève, 1964b: 156. — Type species: *Helmis mixta* Grouvelle, 1899. — Gender feminine.

mixta (Grouvelle, 1899)

Helmis mixta Grouvelle, 1899: 185

Type Locality: South Africa.

Distribution: South Africa.

Protelmis Grouvelle, 1911c

Protelmis Grouvelle, 1911c: 265. — Type species: *Protelmis limnioides* Grouvelle, 1911c. — Gender feminine.

chutteri Delève, 1967c

Protelmis chutteri Delève, 1967c: 83

Type Locality: Zimbabwe.

Distribution: South Africa, Zimbabwe.

limnioides Grouvelle, 1911c

Protelmis limnioides Grouvelle, 1911c: 266

Type Locality: Uganda.

Distribution: Uganda.

Note: The figure captions in the original description are partly wrong, the habitus of this species is shown in Grouvelle (1911c: pl. 1: fig. 7 (not fig. 3)).

propinqua Grouvelle, 1920

Protelmis propinqua Grouvelle, 1920: 205

TYPE LOCALITY: Kenya.

DISTRIBUTION: Angola, Congo (R), Ghana, Kenya.

Troglelmis Jeannel, 1950

Troglelmis Jeannel, 1950: 168. — Type species: *Troglelmis leleupi* Jeannel, 1950. — Gender feminine.

leleupi Jeannel, 1950

Troglelmis leleupi Jeannel, 1950: 171

TYPE LOCALITY: Congo (DR).

DISTRIBUTION: Congo (DR).

References

Abeille de Perrin, E., 1900. Esolus galloprovincialis, p. 137. In Abeille de Perrin, E. & A. Grouvelle (eds.), Descriptions de deux Elmides nouveaux de France (Col.). Bulletin de la Société Entomologique de France 1900 (6): 137.

Aleksandrova, G.N. & N.I. Zaporozhets, 2008. Palynological characteristics of Upper Cretaceous and Paleogene deposits on the west of the Sambian peninsula (Kaliningrad Region), Part 2. Stratigraphiya, Geologicheskaya Korrelyatiya 16 (5): 75–86.

Allard, E., 1869. Description de quelques Coléoptères nouveaux et notes diverses. L'Abeille, Mémoires d'Entomologie 5 (1868–1869): 465–478.

Alluaud, C., 1922. Les Helmides du Nord de l'Afrique. Descriptions d'espèces nouvelles du Maroc (Insectes Coléoptères). Bulletin de la Société des Sciences Naturelles du Maroc 2 (1–2): 31–43.

Alluaud, C., 1926. Compte rendu d'une mission zoologique dans le Maroc sud-oriental (Avril–Mai 1924). Bulletin de la Société des Sciences Naturelles du Maroc 6 (1–6): 12–28, 1 map.

Alluaud, C., 1933. Voyage de Ch. Alluaud et P.A. Chappuis en Afrique occidentale française (Décembre 1930.[sic]–Avril 1931). Helminthidæ [Coléopt.]. Annales de la Société Entomologique de France 102: 155–160.

Angell, G.W.J., 1892. Two new species of Coleoptera. Entomological News 3 (4): 84.

Anonymous [Broun, T.], 1882. Alteration of generic names. The Annals and Magazine of Natural History (Ser. V) IX (53): 409.

Archangelsky, M. & C. Brand, 2014. A new species of Luchoelmis Spangler & Staines (Coleoptera: Elmidae) from Argentina and its probable larva. Zootaxa 3779 (5): 563–572.

Babington, C.C., 1832. Additions to the List of British Insects. The Magazine of Natural History 5: 327–330.

Bameul, F., 1996. Les Hydrethus Fairmaire (Coleoptera, Elmidae). Bulletin de la Société Entomologique de France 101 (3): 273–288.

Barbosa, F.F., A.S. Fernandes & L.G. Oliveira, 2013. Three new species of Macrelmis Motschulsky, 1859 (Coleoptera: Elmidae: Elminae) from the Brazilian Cerrado Biome with updated key for the Macrelmis of Brazil. Zootaxa 3736 (2): 128–142.

Barr, C.B., 1984. Dubiraphia harleyi, a new species of Riffle Beetle from Louisiana (Coleoptera: Dryopoidea: Elmidae). Journal of the Kansas Entomological Society 57 (2): 336–339.

Barr, C.B., 2011. Bryelmis Barr (Coleoptera: Elmidae: Elminae), a new genus of Riffle Beetle with three new species from the Pacific Northwest, U.S.A. The Coleopterists Bulletin 65 (3): 197–212.

Bedel, L., 1878. Diverses observations relatives à l'insecte décrit par Latreille sous le nom d'Elmis Maugetii. Annales de la Société Entomologique de France (Ser. 5) 8, Bulletin des Séances de la Società Entomologique de France (5): LXXIV–LXXV.

Berthélemy, C., 1962. Contribution à l'étude systématique des Elminthidae (Coléoptères). Bulletin de la Société d'Histoire Naturelle de Toulouse 97 (1–2): 201–225.

Berthélemy, C., 1964a. Elminthidae d'Europe occidentale et méridionale et d'Afrique du nord [Coléoptères]. Bulletin de la Société d'Histoire Naturelle de Toulouse 99 (1–2): 244–285.

Berthélemy, C., 1964b. Les Elminthidae décrits par C. Rey: note synonymique [Coléoptères Dryopoidea]. Bulletin de la Société d'Histoire Naturelle de Toulouse 99 (3–4): 525–528.

Berthélemy, C., 1979. Elmidae de la région paléarctique occidentale: systématique et répartition (Coleoptera Dryopoidea). Annales de Limnologie 15 (1): 1–102.

Berthélemy, C., 1980. Oulimnius cyneticus, p. 421. In Berthélemy, C. & L.S. Whytton da Terra (eds.), 1980, Hydraenidae et Elmidae du Portugal. Deuxième note (Coleoptera). Bulletin de la Société d'Histoire Naturelle de Toulouse 115 (3–4) [1979]: 414–424.

Berthélemy, C. & F. Clavel, 1961. Répartition des Coléoptères dans un cours d'eau de la bordure occidentale du Massif Central français. Bulletin de la Société d'Histoire Naturelle de Toulouse 96 (3–4): 241–249.

Bertolini, S., 1874. Catalogo sinonimico e topografico dei coleotteri d'Italia. Tipografia Cenniniana, Firenze: 93–156 [according to Dalla Torre (1886) published in parts as supplement to 'Bullettino della Società Entomologica Italiana' 1872–1894].

Bertolini, S., 1900. Catalogo dei coleotteri d'Italia. Rivista italiana di scienze naturali (ed.), Siena: 25–32 [published in parts as supplement to 'Rivista italiana di scienze naturali' 1899–1904].

Bertrand, H., 1972. Larves et nymphes des Coléoptères aquatiques du globe. F. Paillart, Paris: 804 pp.

Bertrand, H. & A.W. Steffan, 1963. Elminthidarum genus novum e regione aethiopica: Pseudancyronyx [Coleoptera: Dryopoidea]. Bulletin de l'Institut français d'Afrique noire (Ser. A) 25 (3): 827–837.

Bian, D., C. Guo & L. Ji, 2012. First record of Ancyronyx Erichson (Coleoptera: Elmidae) from China, with description of a new species. Zootaxa 3255: 57–61.

Bian, D. & L. Ji, 2010. Two new species of Cuspidevia Jäch & Boukal, 1995 from China (Coleoptera: Elmidae). Zootaxa 2663: 53–58.

Blackburn, T., 1894. Notes on Australian Coleoptera with descriptions of new species. Part XV. Proceedings of the Linnean Society of New South Wales (Ser. 2) 9 (1): 85–108.

Blanchard, E., 1841. Famille des Elmiens, pp. 60–61. In d'Orbigny, A., E. Blanchard & A. Brullé (eds.), Insectes de l'Amérique Méridionale (= Voyage dans l'Amérique

Méridionale [...]. Vol. 6 (2), 1837–1845). Bertrand/Levrault, Paris/Strasbourg: 222 pp., pls. 1–32 [publication date confirmed by Sherborn & Griffin (1934: 132)].

Blatchley, W.S., 1910. An illustrated descriptive catalogue of the Coleoptera or beetles (exclusive of the Rhynchophora) known to occur in Indiana. With bibliography and descriptions of new species. The Nature Publishing Co., Indianapolis: 1386 pp.

Blatchley, W.S., 1925. Notes on the distribution and habits of some Florida Coleoptera with descriptions of new species. The Canadian Entomologist 57 (7): 160–168.

Bocak, L., C. Barton, A. Crampton-Platt, D. Chesters, D. Ahrens & A.P. Vogler, 2014. Building the Coleoptera tree-of-life for >8000 species: composition of public DNA data and fit with Linnaean classification. Systematic Entomology 39 (1): 97–110.

Boheman, C.H., 1851. Insecta Caffrariæ annis 1838–1845 a J.A. Wahlberg collecta. Vol. I (II). Officina Norstedtiana, Holmiae: 299–626, 2 pls.

Bold, T.J., 1871. Note on the Hydrochus parumoculatus of Hardy. The Entomologist's Monthly Magazine 7: 35.

Bollow, H., 1938. Monographie der palaearktischen Dryopidae, mit Berücksichtigung der eventuell transgredierenden Arten (Col.). Mitteilungen der Münchner Entomologischen Gesellschaft 28 (2): 147–187.

Bollow, H., 1939. Eine neue Dryopiden-Gattung aus Abessinien. Atti del Museo Civico di Storia Naturale di Trieste 14 (10) [1937–1941]: 149–152.

Bollow, H., 1940a. Die erste Helminide [sic] (Col. Dryop.) aus Bernstein. Mitteilungen der Münchner Entomologischen Gesellschaft 30 (1): 117–119.

Bollow, H., 1940b. Entomological results from the Swedish expedition 1934 to Burma and British India. Coleoptera: Dryopidae, gesammelt von René Malaise. Arkiv för Zoologi 32A (13): 1–37.

Bollow, H., 1941. Monographie der palaearktischen Dryopidae mit Berücksichtigung der eventuell transgredierenden Arten. (Col.) [Fortsetzung]. Mitteilungen der Münchner Entomologischen Gesellschaft 31 (1): 1–88, 1 pl.

Bollow, H., 1942. Eine neue Stenelmis-Art aus Neu-Guinea (Col. Dryopidae). Annales historico-naturales Musei nationalis hungarici, Pars Zoologica 35: 197–200.

Bosse, L.S., D.W. Tuff & H.P. Brown, 1988. A new species of Heterelmis from Texas (Coleoptera: Elmidae). The Southwestern Naturalist 33 (2): 199–203.

Bouchard, P., Y. Bousquet, A.E. Davies, M.A. Alonso-Zarazaga, J.F. Lawrence, C.H.C. Lyal, A.F. Newton, C.A.M. Reid, M. Schmitt, S.A. Ślipiński & A.B.T. Smith, 2011. Family-group names in Coleoptera (Insecta). ZooKeys 88: 1–972.

Boukal, D.S., 1997. A revision of the genus Austrolimnius Carter & Zeck, 1929 (Insecta: Coleoptera: Elmidae) from New Guinea and the Moluccas. Annalen des Naturhistorischen Museums in Wien (Ser. B) 99: 155–215.

Boumaiza, M., 1994. Recherches sur les eaux courantes de la Tunisie. Faunistique, Ecologie et Biogéographie. Thése de doctorat d'état, Faculté des Sciences, Tunis: 429 pp.

Brancsik, C., 1914. Coleoptera nova. A Trencsénvármegyei Muzeum-Egyesület Értesitője [Bericht des Museumvereines für das Comitat Trencsén]: 58–69.

Broun, T., 1881. Manual of the New Zealand Coleoptera, Part II. Colonial Museum and Geographical Survey Department, Wellington: 653–744 + xxi–xxiii.

Broun, T., 1882. Change of Nomenclature of N.Z. Beetles. The New Zealand Journal of Science 1 (3): 128.

Broun, T., 1885. Abstract of paper on New Zealand Scydmaenidae. The New Zealand Journal of Science 2 (8): 384–387.

Broun, T., 1914. Descriptions of new genera and species of Coleoptera. Bulletin of the New Zealand Institute 1 (3): 143–266.

Brown, H.P., 1970a. Neocylloepus, a new genus from Texas and Central America (Coleoptera: Dryopoidea: Elmidae). The Coleopterists' Bulletin 24 (1): 1–28.

Brown, H.P., 1970b. Neotropical dryopids I. Xenelmis laura, a new species from Brazil (Coleoptera, Elmidae). The Coleopterists' Bulletin 24 (3): 62–65.

Brown, H.P., 1971. A new species of Elsianus from Texas and Mexico, with records of other species in the United States (Coleoptera: Dryopoidea: Elmidae). The Coleopterists Bulletin 25 (2): 55–58.

Brown, H.P., 1972a. Aquatic dryopoid beetles (Coleoptera) of the United States. Biota of Freshwater Ecosystems. Identification Manual No. 6. Water Pollution Conference Series. United States Environmental Protection Agency, Washington, D.C.: ix + 82 pp.

Brown, H.P., 1972b. Synopsis of the genus Heterelmis Sharp in the United States, with description of a new species from Arizona (Coleoptera, Dryopoidea, Elmidae). Entomological News 83 (9): 229–238.

Brown, H.P., 1981a. A distributional survey of the world genera of aquatic dryopoid beetles (Coleoptera: Dryopidae, Elmidae, and Psephenidae sens. lat.). The Pan-Pacific Entomologist 57 (1): 133–148.

Brown, H.P., 1981b. Huleechius, a new genus of riffle beetles from Mexico and Arizona (Coleoptera, Dryopoidea, Elmidae). The Pan-Pacific Entomologist 57 (1): 228–244.

Brown, H.P., 1981c. Key to the world genera of Larinae (Coleoptera, Dryopoidea, Elmidae), with descriptions of new genera from Hispaniola, Colombia, Australia, and New Guinea. The Pan-Pacific Entomologist 57 (1): 76–104.

Brown, H.P., 1983. A catalog of the Coleoptera of America north of Mexico. Family: Elmidae. Agriculture Handbook 529-50, United States Department of Agriculture, Washington, D.C.: 23 pp.

Brown, H.P., 1984. Neotropical dryopoids, III. Major nomenclatural changes affecting Elsianus Sharp and Macrelmis Motschulsky, with checklists of species (Coleoptera: Elmidae: Elminae). The Coleopterists Bulletin 38 (2): 121–129.

Brown, H.P., 1985. Xenelmis sandersoni, a new species of riffle beetle from Arizona and northern Mexico (Coleoptera: Dryopoidea: Elmidae). The Southwestern Naturalist 30 (1): 53–57.

Brown, H.P., 1987. Stenelmis cheryl: new name for a well-known riffle beetle (Coleoptera: Elmidae). Entomological News 98 (3): 111–112.

Brown, H.P., 2001. Synopsis of the riffle beetle genus Zaitzevia (Coleoptera: Elmidae) in North America, with description of a new subgenus and species. Entomological News 112 (3): 201–211.

Brown, H.P. & M.P. Thobias, 1984. World synopsis of the riffle beetle genus Leptelmis Sharp, 1888, with a key to Asiatic species and description of a new species from India (Coleoptera, Dryopoidea, Elmidae). The Pan-Pacific Entomologist 60 (1): 23–29.

Brown, W.J., 1930a. Coleoptera of the north shore of the Gulf of the St. Lawrence. The Canadian Entomologist 62 (2): 239–246.

Brown, W.J., 1930b. New species of Coleoptera I. The Canadian Entomologist 62 (4): 87–92.

Brown, W.J., 1933. New species of Coleoptera IV. The Canadian Entomologist 65 (2): 43–47.

Bug, C., 1973. Zur Genitalmorphologie und Systematik der neotropischen Spezies des Elminthidae-Genus Microcylloepus Hinton, 1935 (Coleoptera: Dryopoidea). Beiträge zur Entomologie 23 (1–4): 99–130.

Bukejs, A., V.I. Alekseev & M.A. Jäch, 2015. The riffle beetles (Coleoptera: Elmidae) of the Eocene Baltic amber: Heterelmis groehni sp. nov. and Heterlimnius samlandicus (Bollow, 1940) comb. nov. Zootaxa 3986 (4): 452–460.

Caillol, H., 1913. Catalogue des Coléoptères de Provence, 2me partie. Société Linnéenne de Provence, Marseille: 607 pp.

Carter, H.J., 1926a. Revision of Athemistus and Microtragus (Fam. Cerambycidae) with notes, and descriptions of other Australian Coleoptera. The Proceedings of the Linnean Society of New South Wales 51 (4): 492–516.

Carter, H.J., 1926b. Revision of the Australasian species of Anilara (Fam. Buprestidae) and Helmis (Fam. Dryopidae), with notes, and descriptions of other Australian Coleoptera. The Proceedings of the Linnean Society of New South Wales 51 (2): 59–71.

Carter, H.J., 1930. Australian Coleoptera. Notes and new species VII. The Proceedings of the Linnean Society of New South Wales 55 (2): 179–190, pl. iv.

Carter, H.J. & E.H. Zeck, 1929. A monograph of the Australian Dryopidae. Order-Coleoptera. The Australian Zoologist 6 (1): 50–72, pls. i–vii.

Carter, H.J. & E.H. Zeck, 1930. Austrolimnius luridus, p. 190. In Carter, H.J. (ed.), Australian Coleoptera. Notes and new species VII. The Proceedings of the Linnean Society of New South Wales 55 (2): 179–190, pl. iv.

Carter, H.J. & E.H. Zeck, 1932. Four new species of Dryopidae, together with notes on the family (Order Coleoptera.). The Australian Zoologist 7 (3): 202–205, pl. xii.

Carter, H.J. & E.H. Zeck, 1933. Three new species of Dryopidae. The Australian Zoologist 7 (4): 371–372, pl. xix.

Carter, H.J. & E.H. Zeck, 1935. Three new species of Dryopidae. The Australian Zoologist 8 (2): 79–80, pl. vii.

Carter, H.J. & E.H. Zeck, 1936. Five new species of Australian Dryopidae. The Australian Zoologist 8 (3): 156–159, pl. xi.

Carter, H.J. & E.H. Zeck, 1938. Four new species of Australian Dryopidae. The Australian Zoologist 9 (2): 170–172, pl. xiv.

Casey, T.L., 1893. Coleopterological Notices. V. The Annals of the New York Academy of Sciences 7: 281–607.

Castelnau, [F.-L.], 1840. Histoire Naturelle des Insectes Coléoptères. Vol. 2. P. Duménil, Paris: 564 pp., 38 pls.

Champion, G.C., 1918. Notes on various South American Coleoptera collected by Charles Darwin during the Voyage of the "Beagle," [sic] with descriptions of new genera and species. The Entomologist's Monthly Magazine 54: 43–55.

Champion, G.C., 1923. Some Indian Coleoptera (11). The Entomologist's Monthly Magazine 59: 165–179.

Champion, G.C., 1927. Some Indian Coleoptera (22). The Entomologist's Monthly Magazine 63: 49–51.

Chandler, H.P., 1949. A new species of Stenelmis from Nevada (Coleoptera, Elmidae). The Pan-Pacific Entomologist 25 (3): 133–136.

Chandler, H.P., 1954. New genera and species of Elmidae (Coleoptera) from California. The Pan-Pacific Entomologist 30 (2): 125–131.

Chavanon, G., A. Berrahou & A. Millan, 2004. Apport à la connaissance des Coléoptères et Hémiptères aquatiques du Maroc Oriental: catalogue faunistique. Boletín de la S.E.A. (Sociedad Entomológica Aragonesa) 35: 143–162.

Chûjô, M. & M. Satô, 1964. 12. Family Elmidae, pp. 195–196. In Kira, T. & T. Umesao (eds.), Nature and Life in Southeast Asia. Vol. III. Fauna and Flora Research Society, Kyoto: VII + 466 pp.

Čiampor, F., 1999. Homalosolus zitnanskae sp.nov. from Borneo (Coleoptera: Elmidae). Entomological Problems 30 (1): 31–34.

Čiampor, F., 2000. Graphelmis prisca sp.nov. from India (Coleoptera: Elmidae). Entomological Problems 31 (2): 183–185.

Čiampor, F., 2001. Systematic revision of the genus Graphelmis (Coleoptera: Elmidae). I. Redescription of the genus and description of four new species. Entomological Problems 32 (1): 17–32.

Čiampor, F., 2002a. Systematic revision of the genus Graphelmis (Coleoptera: Elmidae). II. Graphelmis bandukanensis species group. Entomological Problems 32 (2): 149–161.

Čiampor, F., 2002b. Systematická revízia rodu Graphelmis Delève, 1968 (Insecta: Coleoptera: Elmidae). Thesis, Institute of Zoology, Slovak Academy of Sciences, Bratislava: 161 pp. + pp. 183–186, pp. 17–32, pp. 149–161 [in English, with Slovak title].

Čiampor, F., 2003. Systematic revision of the genus Graphelmis (Coleoptera: Elmidae). III. Graphelmis labralis species group. Entomological Problems 33 (1–2): 31–44.

Čiampor, F., 2004a. A new species of the genus Graphosolus from Tioman Island (Coleoptera: Elmidae). Aquatic Insects 26 (3–4): 221–226.

Čiampor, F., 2004b. Systematic revision of the genus Graphelmis (Coleoptera: Elmidae). IV. Graphelmis scapularis and Graphelmis clermonti species groups. Entomological Problems 34 (1–2): 1–20.

Čiampor, F., 2005a. Systematic revision of the genus Graphelmis (Coleoptera: Elmidae). VI. Graphelmis marshalli species group. Entomological Problems 35 (1): 11–38.

Čiampor, F., 2005b. Systematic revision of the genus Graphelmis (Coleoptera: Elmidae). VII. Graphelmis obesa species group. Entomological Problems 35 (2): 117–122.

Čiampor, F., 2006. Systematic revision of the genus Graphelmis (Coleoptera: Elmidae). VIII. Three new species from the Graphelmis marshalli species group and distributional notes on G. grouvellei. Entomological Problems 36 (1): 13–20.

Čiampor, F. & Z. Čiamporová-Zaťovičová, 2008. A new species of Hedyselmis Hinton and notes on the phylogeny of the genus (Coleoptera: Elmidae). Zootaxa 1781: 55–62.

Čiampor, F., Z. Čiamporová-Zaťovičová, M.A. Jäch & J. Kodada, in prep. Pemonia, a new genus of Protelmidae Jeannel, 1950 (Coleoptera: Dryopoidea), new rank confirmed by molecular analysis, with description of six new species from South America.

Čiampor, F., Z. Čiamporová-Zaťovičová & J. Kodada, 2012. Malaysian species of Dryopomorphus Hinton, 1936 (Insecta: Coleoptera: Elmidae). Zootaxa 3564: 1–16.

Čiampor, F. & J. Kodada, 1998. Elmidae: I. Taxonomic revision of the genus Macronychus Müller (Coleoptera), pp. 219–287. In Jäch, M.A. & L. Ji (eds.), Water Beetles of China. Vol. II. Zoologisch-Botanische Gesellschaft und Wiener Coleopterologenverein, Wien: 371 pp.

Čiampor, F. & J. Kodada, 1999. Description of two new species of the genus Jolyelmis from Mount Roraima, Venezuela (Coleoptera: Elmidae). Entomological Problems 30 (2): 55–60.

Čiampor, F. & J. Kodada, 2004. Systematic revision of the genus Graphelmis (Coleoptera: Elmidae). V. Graphelmis picta species group. Entomological Problems 34 (1–2): 55–102.

Čiampor, F. & J. Kodada, 2006. Dryopomorphus hendrichi sp.nov. from West Malaysia (Coleoptera: Elmidae). Entomological Problems 36 (2): 71–73.

Čiampor, F. & J. Kodada, 2010. Taxonomy of the Oulimnius tuberculatus species group (Coleoptera: Elmidae) based on molecular and morphological data. Zootaxa 2670: 59–68.

Čiampor, F., K. Laššová & Z. Čiamporová-Zaťovičová, 2013. Hypsilara breweri n.sp. from Venezuela: description of new species with notes on the morphology and phylogenetic relationships of the genus (Coleoptera: Elmidae: Larainae). Zootaxa 3635 (5): 591–597.

Čiampor, F. & I. Ribera, 2006. Hedyselmis opis: Description of the larva and its phylogenetic relation to Graphelmis (Coleoptera: Elmidae: Elminae). European Journal of Entomology 103: 627–636.

Collier, J.E., 1972. Optioservus ozarkensis, Optioservus sandersoni, pp. 17–19. In Brown, H.P. (ed.), Aquatic dryopoid beetles (Coleoptera) of the United States. Biota of Freshwater Ecosystems. Identification Manual No. 6. Water Pollution Conference Series. United States Environmental Protection Agency, Washington, D.C.: ix + 82 pp.

Common, I.F.B., 1960. A revision of the Australian stem borers hitherto referred to Schoenobius and Scirpophaga (Lepidoptera: Pyralidae, Schoenobiinae). Australian Journal of Zoology 8 (2): 307–347, 2 pls.

Coquerel, C., 1851. Monographie du genre Potamophilus. Revue et Magasin de Zoologie pure et appliquée (Ser. 2) 3: 591–603, pl. 15.

Cowan, C.F., 1971. On Guérin's Iconographie: particularly the insects. Journal of the Society for the Bibliography of Natural History 6 (1): 18–29.

Crotch, G.R., 1872. Zoology. Part 2. List of the coleoptera found during the progress of the survey, pp. 263–268. In Wilson, C.W. & H.S. Palmer (eds.), Ordnance survey of the Peninsula of Sinai. Vol. 1. Ordnance Survey Office, Southampton [1869]: 323 pp.

Crotch, G.R., 1873. Check list of the Coleoptera of America, north of Mexico. Naturalists' Agency, Salem: 136 pp.

Crowson, R.A., 1955. The natural classification of the families of Coleoptera. Nathaniel Lloyd, London: [2] + 187 pp.

Crowson, R.A., 1981. The Biology of the Coleoptera. Academic Press, London: v–xii + 802 pp.

Curtis, J., 1830. British Entomology. Vol. 7. J. Curtis, London: pls. 290–337 (+2 pp. per pl.) + 4 unnumbered pp.

Dalla Torre, K.W., 1886. Die zoologische Literatur von Tirol und Vorarlberg (bis inclusive 1885). Zeitschrift des Ferdinandeums für Tirol und Vorarlberg (3) 30: 321–407.

Darlington, P.J., 1927. Four new Helmidæ from Cuba, with notes on other West Indian species. Psyche 34 (2): 91–97.

Darlington, P.J., 1928. New Coleoptera from western hot springs. Psyche 35 (1): 1–6.

Darlington, P.J., 1929. On the dryopid beetle genus Lara. Psyche 36 (4): 328–331.

Darlington, P.J., 1936. A list of the West Indian Dryopidæ (Coleoptera), with a new genus and eight new species, including one from Colombia. Psyche 43 (2–3): 65–83.

Dejean, [P.F.M.A.], 1821. Catalogue de la collection des Coléoptères de M. le Baron Dejean. Crevot, Paris: viii + 136 + [2] pp.

Dejean, [P.F.M.A.], 1833. Catalogue des Coléoptères de la collection de M. le Comte Dejean. Second edition. Méquignon-Marvis Père et Fils, Paris [1833–1836]: 443 pp. [only pp. 1–176 (livraisons 1–2) issued in 1833].

Dejean, [P.F.M.A.], 1836. Catalogue des Coléoptères de la collection de M. le Comte Dejean. Third edition. Méquignon-Marvis Père et Fils, Paris [1836–1837]: xiv + 503 pp. [only pp. 1–384 (livraisons 1–4) issued in 1836].

Delève, J., 1937a. Dryopidae du Congo belge. Bulletin & Annales de la Société Entomologique de Belgique 77: 149–164.

Delève, J., 1937b. Dryopidae du Congo Belge. I. Larinae. Revue de Zoologie et de Botanique Africaines 30 (1): 88–106.

Delève, J., 1938. Dryopidae du Congo Belge II. Dryopinae. – III. Elminae. Revue de Zoologie et de Botanique Africaines 31 (3–4): 351–375.

Delève, J., 1942. Contribution à l'étude des Dryopidae. [I. Deux Leptelmis nouveaux du Congo belge. (1–5), II. Remarques sur le genre Microdinodes et description d'une espèce nouvelle (5–8)]. Bulletin du Musée royal d'Histoire naturelle de Belgique 18 (59): 1–8.

Delève, J., 1945a. Contribution à l'étude des Dryopidae. III. Le genre Pseudomacronychus Grouvelle, et le dimorphisme alaire de ses espèces. Bulletin du Musée royal d'Histoire naturelle de Belgique 21 (9): 1–12.

Delève, J., 1945b. Contribution à l'étude des Dryopidae. IV. Les espèces du genre Helminthopsis Grouvelle. Bulletin du Musée royal d'Histoire naturelle de Belgique 21 (10): 1–12.

Delève, J., 1945c. Notes sur quelques Dryopidae (Col.) du Congo Belge et description d'espèces nouvelles. Bulletin & Annales de la Société Entomologique de Belgique 81: 149–156.

Delève, J., 1946. Coleoptera XII. Dryopidae, pp. 323–330. In Jeannel, R. (ed.), Mission Scientifique de l'Omo 6 (59). Mémoires du Muséum National d'Histoire Naturelle (Nouvelle Série) 19 (2) [1945].

Delève, J., 1954. Contribution à l'étude des Dryopidae de Madagascar. Le Naturaliste Malgache 6 (1–2): 25–34.

Delève, J., 1955. Dryopidæ (Coleoptera Macrodactyla), pp. 17–22. In Parc National de l'Upemba. I. Mission G.F. de Witte en collaboration avec W. Adam, A. Janssens, L. van Meel et R. Verheyen (1946–1949) 35 (2). Institut des Parcs Nationaux du Congo Belge, Bruxelles.

Delève, J., 1956. Contributions à l'étude de la faune entomologique du Ruanda-Urundi (Mission P. Basilewsky 1953). XCIII. Coleoptera Dryopidae. Annales du Musée royal du Congo Belge Tervuren (Ser. in 8°) Zoologie 51: 374–383.

Delève, J., 1962. Contribution à l'étude des Dryopoidea. I. – Les espèces malgaches du genre Potamodytes Grouvelle (Coleoptera Elminthidae). Bulletin de l'Institut Royal des Sciences Naturelles de Belgique 38 (28): 1–12.

Delève, J., 1963a. Contribution à l'étude des Dryopoidea. II. – Les espèces malgaches du genre Pachyelmis Fairmaire (Coleoptera Elminthidae). Désignation de lectotypes et descriptions d'espèces nouvelles. Bulletin de l'Institut Royal des Sciences Naturelles de Belgique 39 (2): 1–21.

Delève, J., 1963b. Contribution à l'étude des Dryopoidea. IV. – Résultats d'une campagne en Anatolie (E. Janssens, 1962). Dryopidae et Elminthidae. Bulletin de l'Institut Royal des Sciences Naturelles de Belgique 39 (16): 1–11.

Delève, J., 1963c. Contribution à l'étude des Dryopoidea.V. Note sur les divisions génériques des Larinae d'Afrique (Coleoptera Elminthidae) et descriptions d'espèces nouvelles. Bulletin & Annales de la Société Royale d'Entomologie de Belgique 99 (30): 431–458.

Delève, J., 1963d. Contribution à l'étude des Dryopoidea. VI. – Revision des Elminthidae attribués au genre Elmidolia Fairmaire (Coleoptera). Bulletin de l'Institut Royal des Sciences Naturelles de Belgique 39 (23): 1–55.

Delève, J., 1963e. Dryopidae et Elminthidae (Coleoptera Cucujoidea), pp. 83–92. In Exploration du Parc National de la Garamba. – Mission H. de Saeger en collaboration avec P. Baert, G. Demoulin, I. Denisoff, J. Martin, M. Micha, A. Noirfalise, P. Schoemaker, G. Troupin et J. Verschuren (1949–1952) 41 (4). Institut des Parcs Nationaux du Congo et du Rwanda, Bruxelles.

Delève, J., 1963f. Results from the Danish Expedition to the French Cameroons (1949–1950) XXI. – Coleoptera : Dryopidae and Elminthidae. Bulletin de l'Institut français d'Afrique noire (Ser. A) 25 (3): 807–826.

Delève, J., 1964a. Contribution à l'étude des Dryopoidea. VII. Elminthidae nouveaux ou peu connus de Madagascar. Bulletin & Annales de la Société Royale d'Entomologie de Belgique 100 (2): 25–54.

Delève, J., 1964b. Contribution à l'étude des Dryopoidea. VIII. Cinq nouveaux genres d'Elminthidae (Coleoptera) de l'Afrique australe. Bulletin & Annales de la Société Royale d'Entomologie de Belgique 100 (12): 155–176.

Delève, J., 1964c. Contribution à l'étude des Dryopoidea. IX. Le genre Pachyelmis Fairmaire. Les espèces du continent africain (Coleoptera Elminthidae). Bulletin & Annales de la Société Royale d'Entomologie de Belgique 100 (18): 237–254.

Delève, J., 1964d. Contribution à l'étude des Dryopoidea (Coleoptera). X. Dryopidae et Elminthidae de l'Afrique australe. Bulletin & Annales de la Société Royale d'Entomologie de Belgique 100 (38): 509–541.

Delève, J., 1965a. Contribution à l'étude des Dryopoidea. XI. – Lathridelmis et Trachelminthopsis, deux genres nouveaux d'Elminthidae (Coleoptera). Revue de Zoologie et de Botanique Africaines 71 (1–2): 105–112.

Delève, J., 1965b. Contribution à l'étude des Dryopoidea. XIII. Notes sur le genre Helminthopsis Grouvelle et descriptions d'espèces nouvelles (Coleoptera Elminthidae). Bulletin de l'Institut Royal des Sciences Naturelles de Belgique 41 (11): 1–31.

Delève, J., 1965c. Contribution à l'étude des Dryopoidea. XV. – De quelques espèces du genre Pseudomacronychus Grouvelle (Coleoptera Elminthidae). Bulletin de l'Institut Royal des Sciences Naturelles de Belgique 41 (22): 1–11, pls. I–III.

Delève, J., 1965d. Revision du genre Microdinodes et description d'un genre nouveau (Coleoptera Elminthidae) [XIIe contribution à l'étude des Dryopoidea]. Annales / Musée Royal de l'Afrique Centrale Tervuren, Sciences zoologiques 140: 1–57.

Delève, J., 1966a. Contribution à l'étude des Dryopoidea. Notes taxonomiques sur les espèces éthiopiennes du genre Stenelmis Dufour (Coleoptera Elminthidae) et description d'espèces nouvelles. Bulletin & Annales de la Société Royale d'Entomologie de Belgique 102 (3): 57–96.

Delève, J., 1966b. Contribution à l'étude des Dryopoidea. XVII. Elminthidae (Coleoptera) de la province du Cap et de l'Angola. Bulletin & Annales de la Société Royale d'Entomologie de Belgique 102 (4): 97–108.

Delève, J., 1966c. Contribution à l'étude des Dryopoidea. XVIII. – Notes complémentaires sur les espèces africaines du genre Leptelmis Sharp et descriptions d'espèces nouvelles (Coleoptera Elminthidae). Bulletin de l'Institut Royal des Sciences Naturelles de Belgique 42 (8): 1–10.

Delève, J., 1966d. Dryopidae et Elminthidae de l'Angola. Publicações culturais da Companhia de Diamantes de Angola 76: 39–62.

Delève, J., 1967a. Contribution à la faune du Congo (Brazzaville). Mission A. Villiers et A. Descarpentries. XLIV. Coléoptères Elminthidae. Bulletin de l'Institut fondamental d'Afrique noire (Ser. A) 29 (1): 318–343.

Delève, J., 1967b. Contribution à l'étude des Dryopoidea (Coleoptera) XIX. Notes diverses et descriptions d'espèces nouvelles. Bulletin & Annales de la Société Royale d'Entomologie de Belgique 103: 414–446.

Delève, J., 1967c. The Scientific Results of the Hungarian Soil Zoological Expedition to the Brazzaville-Congo. 17. Espèces des familles Dryopidae et Elminthidae (Coleoptera). Opuscula Zoologica (Budapest) 7 (1): 71–85.

Delève, J., 1968a. IV. ~ Coleoptera Elminthidae, pp. 209–272. In Leleup, N. & J. Leleup (eds.), Mission zoologique belge aux îles Galapagos et en Ecuador (N. et J. Leleup, 1964–1965). Résultats scientifiques. Vol. 1. Musée Royal de l'Afrique Centrale & Institut Royal des Sciences naturelles de Belgique.

Delève, J., 1968b. Contributions à la connaissance de la faune entomologique de la Côte-d'Ivoire (J. Decelle, 1961–1964). XIV. Coleoptera Elminthidae. Annales / Musée Royal de l'Afrique Centrale Tervuren, Sciences zoologiques 165: 197–208.

Delève, J., 1968c. Dryopidae et Elminthidae (Coleoptera) du Vietnam. Annales historico-naturales Musei nationalis hungarici, Pars Zoologica 60: 149–181.

Delève, J., 1970a. Chapter VII. Coleoptera: Dryopidae et Elminthidae, pp. 216–227. In Hanström, B., P. Brinck & G. Rudebeck (eds.), South African Animal Life. Results of the Lund University Expedition in 1950–1951. Vol. XIV. Statens naturvetenskapliga forskningsråd, Stockholm.

Delève, J., 1970b. Contribution à la faune de l'Iran. 19. Coléoptères Dryopoidea. Annales de la Société Entomologique de France (N.S.) 6 (3): 701–703.

Delève, J., 1970c. Contribution à l'étude des Dryopoidea. XX. Espèces d'Elminthidae (Coleoptera) peu ou mal connues de l'Amérique du Sud. Bulletin & Annales de la Société Royale d'Entomologie de Belgique 106: 47–67.

Delève, J., 1970d. Contribution à l'étude des Dryopoidea. XXI. Elminthidae (Coleoptera) peu ou mal connus de l'Indonésie et du Vietnam. Bulletin & Annales de la Société Royale d'Entomologie de Belgique 106: 235–273.

Delève, J., 1970e. Contribution à l'étude des Dryopoidea (Coleoptera). XXII. Tableau des espéces du genre Pseudancyronyx Bertrand et Steffan et description d'une espèce et d'une sous-espèce nouvelles. Revue de zoologie et de botanique africaines 82 (3–4): 335–342.

Delève, J., 1970f. Grouvellinus nepalensis n. sp. du Nepal (Coleoptera Elminthidae). Berichte des naturwissenschaftlich-medizinischen Vereins in Innsbruck 58: 319–322.

Delève, J., 1972. Coléoptères Elminthidae récoltés au Gabon par A. Villiers et J. de Muizon. Bulletin de l'Institut fondamental d'Afrique noire (Ser. A) 34 (4): 916–923.

Delève, J., 1973a. Coleoptera: Dryopidae et Elminthidae. Report No. 25 from the Lund University Ceylon Expedition in 1962 (Per Brinck, Hugo Andersson, Lennart Cederholm). Entomologica Scandinavica Supplement 4: 5–23.

Delève, J., 1973b. Elminthidae of Liberia (Coleoptera). Proceedings of the Biological Society of Washington 86 (24): 299–314.

Delève, J., 1973c. Limnichidae, Dryopidae et Elminthidae des Iles Philippines et de l'Archipel Bismarck (Insecta, Coleoptera, Dryopoidea). Steenstrupia 3: 17–30.

Delève, J., 1973d. Results of the Austrian-Ceylonese Hydrobiological Mission, 1970 of the 1st Zoological Institute of the University of Vienna (Austria) and the Department of Zoology of the Vidyalankara University of Ceylon, Kelaniya. Part VII: Dryopidae and Elminthidae of Ceylon. Bulletin of the Fisheries Research Station, Sri Lanka (Ceylon) 24 (1–2): 69–74.

Delève, J., 1974. Elminthidae (Coleoptera) du Ghana. Acta Zoologica Academiae Scientiarum Hungaricae 20 (3–4): 271–288.

Dohrn, C.A., 1882. Exotisches. Entomologische Zeitung [Stettin] 43 (4–6): 245–259.

Drapiez, P.A.J., 1819. Description de huit espèces d'insectes nouveaux. Annales Générales des Sciences Physiques 1: 45–55.

Dubois, A., Crochet, P. A., Dickinson, E. C., Nemésio, A., Aescht, E., Bauer, A. M., Blagoderov, V., Bour, R., De Carvalho, M. R., Desutter-Grandcolas, L., Frétey, T., Jäger, P., Koyamba, V., Lavilla, E. O., Löbl, I., Louchart, A., Malécot, V., Schatz, H. & A. Ohler, 2013. Nomenclatural and taxonomic problems related to the electronic publication of new nomina and nomenclatural acts in zoology, with brief comments on optical discs and on the situation in botany. Zootaxa, 3735 (1): 1–94.

Dufour, L., 1834. Recherches anatomiques et considérations entomologiques sur quelques insectes coléoptères, compris dans les familles des Dermestins, des Byrrhiens, des Acanthopodes, et des Leptodactyles. Annales des Sciences Naturelles (Ser. 2) 1 (Zoologie): 56–84, pls. 2–4.

Dufour, L., 1835. Recherches anatomiques et considérations entomologiques sur les insectes coléoptères des genres Macronique et Elmis. Annales des Sciences Naturelles (Ser. 2) 3 (Zoologie): 151–174, pls. 6–7.

Dufour, L., 1843. Excursion entomologique dans les montagnes de la vallée d'Ossau. Bulletin de la Société des Sciences, Lettres et Arts, de Pau 1843: 5–118.

Duftschmid, K., 1805. Fauna Austriæ, oder Beschreibung der österreichischen Insecten für angehende Freunde der Entomologie. Vol. 1. Verlag der k.k. priv. akademischen Kunst-, Musik- u. Buchhandlung, Linz, Leipzig: xxxvi + 311 pp.

Erichson, W.F., 1847. Naturgeschichte der Insecten Deutschlands. Erste Abtheilung, Coleoptera. Vol. 3 (4). Nicolaische Buchhandlung, Berlin [1848]: vi + [1] + 968 pp. [parts of book published between 1845 and 1848; pp. 481–800 issued in 1847].

Evenhuis, N.L., 2003. Publication and dating of the journals forming the Annals and Magazine of Natural History and the Journal of Natural History. Zootaxa 385: 1–68.

Evenhuis, N.L., 2012. François-Louis Comte de Castelnau (1802–1880) and the mysterious disappearance of his original insect collection. Zootaxa 3168: 53–63.

Fabricius, J.C., 1792. Entomologia systematica emendata et aucta. Secundum classes, ordines, genera, species adjectis synonimis, locis, observationibus, descriptionibus. Vol. 1 [I]. C.G. Proft, Hafniae: XX + 330 pp.

Fabricius, J.C., 1793. Entomologia systematica emendata et aucta. Vol. 2. C.G. Proft, Hafniae: VIII + 519 pp.

Fairmaire, L., 1863. 74. Elmis subparallelus, pp. 74–75 ["Matériaux pour servir ..."]. In Grenier, A. et al. (eds.), Catalogue des Coléoptères de France et matériaux pour servir à la faune des Coléoptères français. Docteur A. Grenier, and Trésorier de la Societé Entomologique de France, Paris: IV + 79 pp. ["Catalogue des ..."] + 135 pp. ["Matériaux pour servir ..."].

Fairmaire, L., 1871. Essai sur les Coléoptères de Barbarie, septième partie. Annales de la Société Entomologique de France (4) 10 [1870]: 369–404.

Fairmaire, L., 1881. [Diagnoses de Coléoptères nouveaux trouvés en Corse et en Sardaigne]. Annales de la Société Entomologique de France (6) 1, Bulletin des Séances de la Societé Entomologique de France (1): XI–XII.

Fairmaire, L., 1884. Descriptions de Coléoptères recueillis par le Baron Bonnaire en Algérie. Bulletin ou Comptes-Rendus des Séances de la Société Entomologique de Belgique (Ser. III) 42 (Annales de la Société Entomologique de Belgique 28): LIX–LXX.

Fairmaire, L., 1889a. Descriptions de Coléoptères de l'Indo-Chine. Annales de la Société Entomologique de France (6) 8 [1888]: 333–378.

Fairmaire, L., 1889b. Diagnoses de Coléoptères Madécasses. Bulletin ou Comptes-Rendus des Séances de la Société Entomologique de Belgique (Ser. III) 115 (Annales de la Société Entomologique de Belgique 33): XC–XCVI.

Fairmaire, L., 1897a. Matériaux pour la faune coléoptérique de la région malgache, 3e note. Annales de la Société Entomologique de Belgique 41: 92–119.

Fairmaire, L., 1897b. Matériaux pour la faune coléoptérique de la région malgache, 4e note. Annales de la Société Entomologique de Belgique 41: 363–406.

Fairmaire, L., 1898a. Matériaux pour la faune coléoptérique de la région malgache, 5e note. Annales de la Société Entomologique de Belgique 42: 222–268.

Fairmaire, L., 1898b. Matériaux pour la faune coléoptérique de la région malgache, 7e note. Annales de la Société Entomologique de Belgique 42: 463–499.

Fairmaire, L., 1902. Matériaux pour la faune coléoptérique de la région malagache, 12e note. Annales de la Société Entomologique de France 71: 325–388.

Fall, H.C., 1901. List of the Coleoptera of southern California, with notes on habits and distribution and descriptions of new species. Occasional Papers of the California Academy of Sciences 8: 1–282.

Fall, H.C., 1907. Descriptions of new species, pp. 218–272. In Fall, H.C. & T.D.A. Cockerell (eds.), The Coleoptera of New Mexico. Transactions of the American Entomological Society 33 (2–3): 145–272.

Fall, H.C., 1925. New species of Helmis (Coleoptera). Journal of the New York Entomological Society 33: 177–181.

Fernandes, A.S., 2010. Taxonomia de Elmidae (Insecta, Coleoptera) do Município de Presidente Figueiredo, Amazonas, Brasil. Instituto Nacional de Pesquisas da Amazônia – INPA Prográma de Pós-Graduação em Entomologia Divisão do Curso de Entomologia – DCEN, Manaus: 125 pp.

Fernandes, A.S. & N. Hamada, 2012. Description and notes on the bionomics of a new species of Potamophilops Grouvelle, 1896 (Coleoptera: Elmidae: Larainae), from the Cerrado biome in Brazil. Journal of Natural History 46 (11–12): 717–727.

Fernandes, A.S., M.I.S. Passos & N. Hamada, 2010a. A new species of Hintonelmis Spangler (Coleoptera: Elmidae: Elminae) from Central Amazonia, Brazil. Zootaxa 2353: 43–48.

Fernandes, A.S., M.I.S. Passos & N. Hamada, 2010b. The genus Portelmis Sanderson, 1935 (Coleoptera: Elmidae: Elminae): first report in Brazil, description of two new Amazonian species and species key for males. Zootaxa 2517: 33–43.

Fernandes, A.S., M.I.S. Passos & N. Hamada, 2011. Stegoelmis Hinton, 1939 (Coleoptera: Elmidae: Elminae) in Brazil: two new species and a key to the Brazilian species. Zootaxa 2921: 56–64.

Flach, C., 1882. Variabilität der Lareynia-Arten. Deutsche Entomologische Zeitschrift 26 (2): 252–253.

Freitag, H., 2008. A new species of Prionosolus Jäch & Kodada from Palawan, Philippines (Coleoptera: Elmidae). Koleopterologische Rundschau 78: 297–303.

Freitag, H., 2012. Ancyronyx jaechi sp.n. from Sri Lanka, the first record of the genus Ancyronyx Erichson, 1847 (Insecta: Coleoptera: Elmidae) from the Indian Subcontinent, and a world checklist of species. Zootaxa 3382: 59–65.

Freitag, H., 2013. Ancyronyx Erichson, 1847 (Coleoptera, Elmidae) from Mindoro, Philippines, with description of the larvae and two new species using DNA sequences for the assignment of the developmental stages. ZooKeys 321: 35–64.

Freitag, H. & M. Balke, 2011. Larvae and a new species of Ancyronyx Erichson, 1847 (Insecta, Coleoptera, Elmidae) from Palawan, Philippines, using DNA sequences for the assignment of the developmental stages. ZooKeys 136: 47–82.

Freitag, H. & M.A. Jäch, 2007. The genus Ancyronyx Erichson, 1847 (Coleoptera, Elmidae) in Palawan and Busuanga, (Philippines) with descriptions of six new species. Zootaxa 1590: 37–59.

Ganglbauer, L., 1904a. Die Käfer von Mitteleuropa. Vol. 4 (1). Karl Gerolds Sohn, Wien: 286 pp.

Ganglbauer, L., 1904b. Neue Arten aus den Gattungen Trechus (Anophthalmus), Hydroporus und Riolus. Münchener Koleopterologische Zeitschrift 2 (2): 350–354.

Gauthier, H., 1928. Recherches sur la faune des eaux continentales de l'Algérie et de la Tunisie. Minerva, Alger: 419 pp., 1 pl., 1 map.

Gemminger, M., 1851. Systematische Uebersicht der Käfer um München. Ein Beitrag zu den Localfaunen Deutschlands. Friedrich Mauke, Jena: X + 65 pp., 1 pl.

Gemminger, [M.] & Harold, [E.] de, 1868. Catalogus Coleopterorum hucusque descriptorum synonymicus et systematicus. Vol. III. E.H. Gummi, Monachii: 978 pp. + unnumbered generic index.

Gerhardt, J., 1869. Die Wasserkäferfauna der weifsen [sic] Wiese im Riesengebirge. Berliner Entomologische Zeitschrift 13: 259–261.

Germain, P., 1854. Elmis chilensis, p. 327. In Germain, P. & R.A. Philippi (eds.), Descripcion de 21 especies nuevas de Coleópteros de Chile.–Preceden algunas observaciones jenerales sobre los Coleópteros de Chile. Anales de la Universidad de Chile 11: 325–336.

Germain, P., 1892. Notes sur les Coléoptères du Chili. Actes de la Société Scientifique du Chili 2: 241–261.

Germar, E.F., 1811. Eine neue Käfergattung Potamophilus. Neue Schriften der naturforschenden Gesellschaft zu Halle 1 (6): 41–46.

Germar, E.F., 1824. Insectorum species novae aut minus cognitae, descriptionibus illustratae. Vol. 1. Coleoptera. Hendel & Sohn, Halle: XXIV + 624 pp., 2 pls.

Gistel, J., 1848. Faunula monacensis cantharologica. Isis von Oken 1848 (6–11): [13 unnumbered pp., inserted between numbered pages (columns) of fascicles 6–11].

Glaister, A., 1999. Guide to the identification of Australian Elmidae larvae (Insecta: Coleoptera). identification Guide No. 21. Cooperative Research Centre for Freshwater Ecology, Thurgoona: 48 pp.

Gómez, E. & J.C. Bello, 2006. Nueva especie de Austrolimnius Carter & Zeck 1929 (Coleoptera: Elmidae: Elminae) de Venezuela, y descripción de su larva. Entomotropica 21 (1): 13–17.

Gory, M., 1844. Potamophilus orientalis, p. 70. In Guérin Méneville, F.E. (ed.), Iconographie du règne animal de G. Cuvier. Insectes. [Vol. 1]. J.B. Baillière, Londres [1829–1838]: 576 pp. [publication date (cited as "1829–1844" by Zaitzev 1910: 7) confirmed by Cowan (1971: 20)].

Gozis, M. des, 1886. Recherche de l'espèce typique de quelques anciens genres. Rectifications synonymiques et notes diverses. Herbin, Montluçon: 36 pp.

Grenier, A., 1863. 74. Catalogue des Coléoptères de France, pp. 1–79. In Grenier, A. et al. (eds.), Catalogue des Coléoptères de France et matériaux pour servir à la faune des Coléoptères français. Docteur A. Grenier, and Trésorier de la Société Entomologique de France, Paris: IV + 79 pp. ["Catalogue des ..."] + 135 pp. ["Matériaux pour servir ..."].

Grouvelle, A., 1889a. [Divers renseignements relatifs au groupe des Helmis et la diagnose d'un Riolus nouveau]. Annales de la Société Entomologique de France (Ser. 6) 9, Bulletin des Séances et Bulletin bibliographique de la Société Entomologique de France (V): LXXIX–LXXX.

Grouvelle, A., 1889b. Nouvelles espèces d'Helmides. Annales de la Société Entomologique de France (Ser. VI) 8 [1888]: 393–410, pls. VII–VIII.

Grouvelle, A., 1889c. Voyage de M.E. Simon au Venezuela (Décembre 1887–Avril 1888). Coléoptères, 2e mémoire (1). Cucujidæ, Rhysodidæ, Dryopidæ, Cyathoceridæ, Heteroceridæ. Annales de la Société Entomologique de France (Ser. VI) 9: 157–166, pl. 6.

Grouvelle, A., 1890. [Diagnoses de trois Helmis nouveaux]. Annales de la Société Entomologique de France (Ser. 6) 10, Bulletin des Séances et Bulletin bibliographique de la Société Entomologique de France (XIV): CCXII.

Grouvelle, A., 1892a. Trois espèces nouvelles d'Helmides des Îles de la Sonde. Notes from the Leyden Museum 14 (3–4): 187–190.

Grouvelle, A., 1892b. Viaggio di Leonardo Fea in Birmania e regioni vicine.L. Nitidulides, Cucujides et Parnides. 2.me Partie. Annali del Museo Civico di Storia Naturale di Genova (Serie 2), 12 (XXXII): 833–868.

Grouvelle, A., 1894. Insectes du Bengale. (35e mémoire). Clavicornes. Annales de la Société Entomologique de Belgique 38 (10): 578–587.

Grouvelle, A., 1895. Voyage de M.E. Simon dans l'Afrique austral (Janvier–Avril 1893), 4e Mémoire. Nitidulidæ, Colydiidæ, Cucujidæ, Cryptophagidæ & Parnidæ. Annales de la Société Entomologique de France 64: 161–168.

Grouvelle, A., 1896a. Descriptions de Dryopides (Parnides) et Helmides nouveaux. Notes from the Leyden Museum 18 (1): 33–52.

Grouvelle, A., 1896b. Note sur les Pelonomus pubescens Blanch. et simplex Berg et description de quelques Dryopides et Helmides nouveaux. Anales del Museo Nacional de Buenos Aires 5: 5–10 [originally published separately on 8.V.1896 with dual page numbers: 1–6/(5–10); reprinted in Anales del Museo Nacional de Buenos Aires "Segunda Serie, Tomo V (Ser. 2[a], t. II)" [1896–1897]: 5–10].

Grouvelle, A., 1896c. Note sur les subdivisions génériques des Potamophiliens (Col.). Bulletin de la Société Entomologique de France 1896, Bulletin des Séances et Bulletin bibliographique de la Société Entomologique de France (4): 77–79.

Grouvelle, A., 1896d. Potamophilides, Dryopides, Helmides et Heterocerides des Indes orientales. Annali del Museo Civico di Storia Naturale di Genova (Serie II), 17 (XXXVII): 32–56.

Grouvelle, A., 1896e. Remarques sur la nomenclature des Dryopides et des Helmides [Col.]. Bulletin de la Société Entomologique de France 1896, Bulletin des Séances et Bulletin bibliographique de la Société Entomologique de France (2): 27.

Grouvelle, A., 1896f. Remarques synonymiques sur les Dryopides, Helmides Hétérocérides et spécialement sur les types de la collection Kuwert [Col.]. Bulletin de la Société Entomologique de France 1896, Bulletin des Séances et Bulletin bibliographique de la Société Entomologique de France (4): 75–77.

Grouvelle, A., 1897. Remarques synonymiques sur quelques types d'Helmides et d'Hétérocerides de la collection du Musée de Sarajevo. Bulletin de la Société Entomologique de France 1897 (13): 206.

Grouvelle, A., 1898a. Clavicornes de Grenada et de S[t]. Vincent (Antilles) récoltés par M.H.H. Smith, et appartenant au Musée de Cambridge. Notes from the Leyden Museum 20 (1): 35–48.

Grouvelle, A., 1898b. Clavicornes nouveaux. Annales de la Société Entomologique de Belgique 42 (3): 110–112.

Grouvelle, A., 1899. Descriptions de Clavicornes d'Afrique et de la région Malgache. 2e Mémoire. Annales de la Société Entomologique de France 68: 136–185.

Grouvelle, A., 1900a. [Insectes de Congo] Clavicornes. Annales de la Société Entomologique de Belgique 44 (11): 424–425.

Grouvelle, A., 1900b. Contribution à l'étude de la faune entomologique de Sumatra (Côte ouest: Indrapoera – Vice-résidence de Païnan) (Voyage de M.J.-L. Weyers). XIIe Article. Nitidulides, Colydiides, Rhysodides, Cucujides, Monotomides, Cryptophagides, Tritomides, Dryopides. Annales de la Société Entomologique de Belgique 44 (7): 262–268.

Grouvelle, A., 1900c. Riolus meridionalis, p. 137. In Abeille de Perrin, E. & A. Grouvelle (eds.), Descriptions de deux Elmides nouveaux de France (Col.). Bulletin de la Société Entomologique de France 1900 (6): 137.

Grouvelle, A., 1902a. Clavicornes nouveaux de Musée Royal de Bruxelles (Afrique et Australie). Annales de la Société Entomologique de Belgique 46 (5): 184–190.

Grouvelle, A., 1902b. Phanocerus Bugnioni, p. 466. In Grouvelle, A. & A. Léveillé (eds.), Voyage de M. le D[r] Ed. Bugnion au Vénézuela, en Colombie et aux Antilles. Coléoptères Clavicornes. Annales de la Société Entomologique de France 71: 461–467.

Grouvelle, A., 1906a. Contribution à l'étude des Coléoptères de Madagascar. Annales de la Société Entomologique de France 75: 67–168, pls. 7–8.

Grouvelle, A., 1906b. Nitidulides, Colydiides, Cucujides, Monotomides et Helmides nouveaux. Revue d'Entomologie 25: 113–126.

Grouvelle, A., 1906c. Voyage de M. Ch. Alluaud dans l'Afrique orientale. Juin 1903 à mai 1904. Dryopidae, Helminthidae, Heteroceridae (Coléoptères). Annales de la Société Entomologique de France 75: 315–332, pl. 10.

Grouvelle, A., 1908. Mission de M.F. Geay dans la Guyane (bassin du fleuve Carsevenne), Coléoptères : Helminthidæ. Bulletin du Muséum national d'histoire naturelle 14 (3): 181–186.

Grouvelle, A., 1911a. Description d'un Microdes nouveau [Col. Dryopidae]. Bulletin de la Société Entomologique de France 1911 (12): 252–253.

Grouvelle, A., 1911b. XIX. Descriptions of five new Indian species of clavicorn Coleoptera. Records of the Indian Museum 1911 (5): 315–317.

Grouvelle, A., 1911c. Voyage de M. Ch. Alluaud en Afrique orientale et centrale. Septembre 1908 à mai 1909. Helminthidae [Coléoptères]. Annales de la Société Entomologique de France 80: 265–282, pl. 1.

Grouvelle, A., 1913. Mission Tilho (Niger-Tchad). Coléoptères Clavicornes. Bulletin du Muséum national d'histoire naturelle 19 (8): 569–573.

Grouvelle, A., 1914. Descriptions de Coléoptères africains. Annales de la Société de France 83: 141–202, pls. 6–7.

Grouvelle, A., 1920. Dryopidae, Georyssidae et Heteroceridae, pp. 193–228, pl. IX ["VIII"]. In Voyage de Ch. Alluaud et R. Jeannel en Afrique orientale (1911–1912). Résultats scientifiques. Coleoptera XV. L. Lhomme, Paris.

Guenée, A., 1857. Uranides et Phalénites. Vol. II. In Boisduval, [J.A.] & [A.] Guenée (eds.), Histoire naturelle des insectes. Species général des Lépidoptères. Vol. 10. Librairie Encyclopédique de Roret, Paris: 584 pp.

Guérin-Méneville, F.E., 1835. Iconographie du règne animal de G. Cuvier. Vol. II (Planches des Animaux invertébrés). J.B. Baillière, Londres [1829–1844]: pls. 15–20, 51 [publication date suggested by Cowan (1971: 29)].

Guérin Méneville, [F.E.], 1843. Potamophilus Goudotii, Potamophilus Cordillieræ, pp. 18–19. In Guérin Méneville, [F.E.] & J. Goudot (eds.), Insectes nouveaux, observés sur les plateaux des Cordillères et dans les vallées chaudes de la Nouvelle-Grenade; avec des notes relatives à leurs mœurs, à leur distribution géographique, etc. Revue zoologique 1843: 12–22.

Guillebeau, F., 1892. [Descriptions de deux Coléoptères nouveaux]. Annales de la Société Entomologique de France 61, Bulletin des Séances et Bulletin bibliographique de la Société Entomologique de France (IX): CXXXIII–CXXXV.

Gutierrez, J., 1969. Tetranychidae nouveaux de Madagascar (Cinquième note). Acarologia 11: 43–64.

Gyllenhal, L., 1808. Insecta Svecica. Classis I. Coleoptera sive Eleuterata. Vol. 1. F.J. Leverentz, Scaris: VIII + 572 pp.

Gyllenhall, L., 1827. Insecta Svecica. Classis I. Coleoptera sive Eleuterata. Vol. 1 (IV). Fleischer, Lipsia: VIII + [2] + 762 pp.

Hansen, M., 1999. Hydrophiloidea (Coleoptera). In Hansen, M. (ed.), World Catalogue of Insects. Vol. 2. Stenstrup: Apollo Books, 416 pp.

Hardy, J., 1854. H.[ydrochus] parumoculatus, p. 270. In Hardy, J. & T.J. Bold (eds.), IX.–A Catalogue of the Insects of Northumberland and Durham (Part iii). Transactions of the Tyneside Naturalists' Field Club 2 [1851–1854]: pp. 164–287.

Hatch, M.H., 1938. Two new species of Helmidae from a warm spring in Montana (Coleoptera). Entomological News 49 (1): 16–19.

Haupt, H., 1956. Beitrag zur Kenntnis der eozänen Arthropodenfauna des Geiseltales. Nova Acta Leopoldina 18 (128): 1–90.

Hay, J.S., 1876. On the District of Akem, in West Africa. Proceedings of the Geographical Society of London 20 (6) [1875–1876]: 475–482.

Hayashi, M. & Y. Okushima, 2012. A fossil record of Elminae (Coleoptera, Elmidae) from the Upper Miocene in Tastumi[= Tatsumi]-tôge, Tottori Prefecture, Western Honshu, Japan. Bulletin of the Kurashiki Museum of natural History 27: 5–7 [in Japanese, English title and abstract].

Hayashi, M., S.D. Song & T. Sota, 2013. Patterns of hind-wing degeneration in Japanese riffle beetles (Coleoptera: Elmidae). European Journal of Entomology 110 (4): 689–697.

Heer, O., 1841. Fauna Coleopterorum Helvetica. Vol. 1. Orelii, Fuesslini et Sociorum, Turicum [1838–1841]: XII + 652 pp.

Herman, L.H., 2001. Catalog of the Staphylinidae (Insecta: Coleoptera). 1758 to the end of the second millennium. VII. Bibliography and index. Bulletin of the American Museum of Natural History 265: 3841–4218.

Hernando, C., P. Aguilera & I. Ribera, 2001. Limnius stygius sp.nov., the first stygobiontic riffle beetle from the Palearctic Region (Coleoptera: Elmidae). Entomological Problems 32 (1): 69–72.

Hernando, C., I. Ribera & P. Aguilera, 1998. Description of the adults and larvae of a remarkable new Oulimnius Gozis from the Anti-Atlas (S.W. Morocco) (Coleoptera: Elmidae). Annales de la Société Entomologique de France (N.S.) 34 (3): 253–258.

Heyden, L. von, 1870. Beschreibungen neuer Arten, pp. 58–176. In Heyden, L. von (ed.), Entomologische Reise nach dem südlichen Spanien, der Sierra Guadarrama und Sierra Morena, Portugal und den Cantabrischen Gebirgen. Entomologischer Verein Berlin, Berlin: 218 pp., pls. I–II.

Heyden, L. von, E. Reitter & J. Weise, 1906. Catalogus coleopterorum Europae, Caucasi et Armeniae rossicae. Editio secunda. R. Friedländer & Sohn, Berlin, Paskau, Caen: 774 pp.

Hilsenhoff, W.L., 1973. Notes on Dubiraphia (Coleoptera: Elmidae) with descriptions of five new species. Annals of the Entomological Society of America 66 (1): 55–61.

Hinton, H.E., 1934. Miscellaneous studies in the Helminae (Dryopidae, Coleoptera). Revista de Entomologia 4 (2): 192–201.

Hinton, H.E., 1935. Notes on the Dryopoidea (Col.). Stylops 4 (8): 169–179.

Hinton, H.E., 1936a. A new genus and a new species of Elminae (Coleoptera, Dryopidae). The Entomologist's Monthly Magazine (3rd Series, 22) 72: 1–5.

Hinton, H.E., 1936b. Descriptions and figures of new Brazilian Dryopidae (Coleoptera). The Entomologist 69 (883): 283–289.

Hinton, H.E., 1936c. Descriptions of new genera and species of Dryopidae (Coleoptera). The Transactions of the Royal Entomological Society of London 85 (18): 415–434, pl. 1.

Hinton, H.E., 1936d. New Dryopidae from the Japan Empire (Coleoptera). The Entomologist 69 (878): 164–168.

Hinton, H.E., 1936e. XVII.–Results of the Oxford University Expedition to Borneo, 1932. Dryopidæ (Coleoptera).–Part II. The Annals and Magazine of Natural History (Ser. 10) xviii (104): 204–224.

Hinton, H.E., 1936f. Synonymical and other notes on the Dryopidae (Coleoptera). The Entomologist's Monthly Magazine (3rd Series, 22) 72: 54–58.

Hinton, H.E., 1937a. Additions to the Neotropical Dryopidae (Coleoptera.). Arbeiten über morphologische und taxonomische Entomologie aus Berlin-Dahlem 4 (2): 93–111.

Hinton, H.E., 1937b. Descriptions and figures of new Peruvian Dryopidae (Coleoptera). The Entomologist 70 (889): 131–138.

Hinton, H.E., 1937c. XXII.–New African Lavinæ [sic] (Coleoptera, Dryopidæ). The Annals and Magazine of Natural History (Ser. 10) xix (110): 289–304.

Hinton, H.E., 1937d. New species of Cylloepus from Brazil (Coleoptera, Dryopidae). The Entomologist 70 (895): 279–284.

Hinton, H.E., 1937e. Notes on some Brazilian Potamophilinae and Elminae (Coleoptera, Dryopidae). The Entomologist's Monthly Magazine (3rd Series, 23) 73: 95–100.

Hinton, H.E., 1939a. An inquiry into the natural classification of the Dryopoidea, based partly on a study of their internal anatomy (Col.). The Transactions of the Royal Entomological Society of London 89 (7): 133–184, pl. 1.

Hinton, H.E., 1939b. A note on the genus Austrolimnius C. & Z., with a description of a new species from French Guiana (Coleoptera, Elmidae). Proceedings of the Royal Entomological Society of London (Ser. B) 8 (10): 195–199.

Hinton, H.E., 1939c. Notes on American Elmidae, with descriptions of new species (Coleoptera). The Entomologist's Monthly Magazine (3rd Series, 25) 75: 179–185.

Hinton, H.E., 1939d. On some new and little known South American Neoelmis Musgrave (Coleoptera, Elmidae). The Entomologist's Monthly Magazine (3rd Series, 25) 75: 228–234.

Hinton, H.E., 1939e. On some new genera and species of neotropical Dryopoidea (Coleoptera). The Transactions of the Royal Entomological Society of London 89 (3): 23–46.

Hinton, H.E., 1940a. A monographic revision of the Mexican water beetles of the family Elmidae. Novitates Zoologicae 42 (2): 217–396.

Hinton, H.E., 1940b. A monograph of Gyrelmis gen.n., with a study of the variation of the internal anatomy (Coleoptera, Elmidae). The Transactions of the Royal Entomological Society of London 90 (13): 375–409.

Hinton, H.E., 1940c. XLI.–A synopsis of the Bolivian species of Cyllœpus Er. (Coleoptera, Elmidæ). The Annals and Magazine of Natural History (Ser. 11) vi (35): 393–409.

Hinton, H.E., 1940d. A synopsis of the Brazilian species of Microcylloepus (Coleoptera, Elmidae). The Entomologist's Monthly Magazine (4th Series, 1) 76: 61–68.

Hinton, H.E., 1940e. XI.–A synopsis of the Brazilian species of Neoelmis Musgrave (Coleoptera, Elmidæ). The Annals and Magazine of Natural History (Ser. 11) v (26): 129–153.

Hinton, H.E., 1940f. A synopsis of the genus Macronychus Müller (Coleoptera, Elmidae). Proceedings of the Royal Entomological Society of London (Ser. B) 9 (7): 113–119.

Hinton, H.E., 1940g. XXVIII.–Entomological expedition to Abyssinia, 1926–7. Coleoptera, Psephenidæ, Dryopidæ, Elmidæ. The Annals and Magazine of Natural History (Ser. 11) vi (33): 297–307.

Hinton, H.E., 1940h. XIX.–On some new Brazilian Microcyllœpus, with a key to the species (Coleoptera, Elmidæ). The Annals and Magazine of Natural History (Ser. 11) vi (32): 236–247.

Hinton, H.E., 1940i. The Percy Sladen Trust Expedition to Lake Titicaca in 1937 under the leadership of Mr H. Cary Gilson. VII. The Peruvian and Bolivian species of Macrelmis Motsch. (Coleoptera, Elmidae). The Transactions of the Linnean Society of London (Ser. 3) 1 (2): 117–147.

Hinton, H.E., 1941a. A synopsis of the American species of Austrolimnius Carter (Col., Elmidae). The Entomologist's Monthly Magazine (4th Series, 2) 77: 156–163.

Hinton, H.E., 1941b. New genera and species of Elmidae (Coleoptera). The Transactions of the Royal Entomological Society of London 91 (3): 65–104.

Hinton, H.E., 1945a. IV.–A synopsis of the Brazilian species of Cyllœpus Er. (Coleoptera, Elmidæ). The Annals and Magazine of Natural History (Ser. 11) xii (85): 43–67.

Hinton, H.E., 1945b. Descriptions of two new species of Elsianus Sharp, with a key to the graniger species-group (Col., Elmidae). The Entomologist's Monthly Magazine (4th Series, 6) 81: 90–92.

Hinton, H.E., 1945c. New and little known species of Microcylloepus (Coleoptera, Elmidae). The Entomologist 78 (983): 57–61.

Hinton, H.E., 1945d. Stethelmis chilensis, new genus and species of Elmidae from Chile (Coleoptera). Proceedings of the Royal Entomological Society of London (Ser. B) 14 (5–6): 73–76.

Hinton, H.E., 1946a. A key to the species of Xenelmis Hinton, with descriptions of three new species (Col., Elmidae). The Entomologist's Monthly Magazine (4th Series, 7) 83: 237–241.

Hinton, H.E., 1946b. A synopsis of the Brazilian species of Elsianus Sharp (Coleoptera, Elmidae). The Transactions of the Royal Entomological Society of London 96 (8): 125–149.

Hinton, H.E., 1946c. LXXII.–A synopsis of the Peruvian species of Cyllœpus Er. (Coleoptera, Elmidæ). The Annals and Magazine of Natural History (Ser. 11) xii (95) [1945]: 713–733 [publication date confirmed by Evenhuis (2003: 50)].

Hinton, H.E., 1948. 11. Coleoptera: Dryopidae and Elmidae, pp. 133–140. In British Museum Expedition to South-West Arabia, 1937/38, 1. Jarrold, Norwich & London.

Hinton, H.E., 1951. LXXVII.–A new Cyllœpus from Argentina (Coleoptera, Elmidæ). The Annals and Magazine of Natural History (Ser. 12) iv (44): 820–823.

Hinton, H.E., 1965. A revision of the Australian species of Austrolimnius (Coleoptera: Elmidae). Australian Journal of Zoology 13 (1): 97–172.

Hinton, H.E., 1968. The subgenera of Austrolimnius (Coleoptera: Elminthidae). The Proceedings of the Royal Entomological Society of London (Ser. B) 37 (7–8): 98–102, pl. 1.

Hinton, H.E., 1969. Structure of the plastron of Hexacylloepus, with a description of a new species (Coleoptera, Elminthidae). Journal of Natural History 3: 125–130.

Hinton, H.E., 1970. The zoological results of Gy. Topál's collectings in south Argentine. 21. A second species of Stethelmis (Coleoptera: Elminthidae). Acta Zoologica Academiae Scientarium Hungaricae 16 (1–2): 109–113.

Hinton, H.E., 1971a. A revision of the genus Hintonelmis Spangler (Coleoptera: Elmidae). The Transactions of the Royal Entomological Society of London 123 (2): 189–208.

Hinton, H.E., 1971b. Pilielmis, a new genus of Elmidae (Coleoptera). The Entomologist's Monthly Magazine (4th Series) 107: 161–166, pls. IV–V.

Hinton, H.E., 1971c. Some American Austrolimnius (Coleoptera: Elmidae). Journal of Entomology (Ser. B) 40 (2): 93–99.

Hinton, H.E., 1971d. The Elmidae (Coleoptera) of Trinidad and Tobago. Bulletin of the British Museum (Natural History), Entomology 26 (6): 245–265, pls. 1–9.

Hinton, H.E., 1971e. The species of Dryopomorphus (Coleoptera, Elmidae). The Entomologist 104 (1302–1303): 293–297, pl. V.

Hinton, H.E., 1972a. Hallazgo de un nuevo Austrolimnius en Guerrero, Mexico (Col., Elmidae). Ciencia 27 (4–5): 135–137.

Hinton, H.E., 1972b. New species of Neoelmis from South America (Coleoptera, Elmidae). Papéis Avulsos de Zoologia 26 (9): 117–135, pls. I–II.

Hinton, H.E., 1972c. The Venezuelan species of Neoelmis (Coleoptera: Elmidae). Journal of Entomology (Ser. B) 41 (2): 133–144.

Hinton, H.E., 1972d. Two new genera of South American Elmidae (Coleoptera). The Coleopterists Bulletin 26 (2): 37–41.

Hinton, H.E., 1973a. New genera and species of Bolivian Elmidae (Coleoptera). The Coleopterists Bulletin 27 (1): 1–6.

Hinton, H.E., 1973b. The Venezuelan species of Hexacylloepus (Col., Elmidae). The Entomologist's Monthly Magazine (4th Series) 108 [1972]: 251–256, pl. VI.

Hinton, H.E., 1976. Hedyselmis, a new genus of Elmidae (Coleoptera) from Malaysia. Systematic Entomology 1: 259–261.

Horn, G.H., 1870. Synopsis of the Parnidae of the United States. Transactions of the American Entomological Society 3: 29–42.

Hua, L., 2002. List of Chinese Insects. Vol. II. Guangzhou Zhongshan University Press, Guangzhou: 612 pp.

Hunt, T., J. Bergsten, Z. Levkanicova, A. Papadopoulou, O.S. John, R. Wild, P.M. Hammond, D. Ahrens, M. Balke, M.S. Caterino, J. Gómez-Zurita, I. Ribera, T.G. Barraclough, M. Bocakova, L. Bocak & A.P. Vogler, 2007. A comprehensive phylogeny of beetles reveals the evolutionary origins of a superradiation. Science 318: 1913–1916.

ICZN, 1988. Opinion 1515. Laridae Rafinesque Schmaltz, 1815 (Aves) and Larini LeConte, 1861 (Insecta, Coleoptera): homonymy removed. The Bulletin of Zoological Nomenclature 45 (3): 245.

ICZN, 1995. Opinion 1812. Elmidae Curtis, 1830 (Insecta, Coleoptera): conserved as the correct original spelling, and the gender of Elmis Latreille, 1802 ruled to be feminine. The Bulletin of Zoological Nomenclature 52 (2): 217–218.

ICZN, 1999. International Code of Zoological Nomenclature. Fourth edition. International Trust for Zoological Nomenclature, London: 306 pp.

Illies, J., 1953. Die deutschen Arten der Dryopidengattung Lathelmis Rttr. Entomologische Blätter 49: 173–184.

Illiger, J., 1802. Aufzählung der Käfergattungen nach der Zahl der Fufsglieder [sic]. Magazin für Insektenkunde 1 (3–4): 285–305.

Jäch, M. [A.], 1982a. Die Koleopterenfauna der Bergbäche von Südwest Ceylon. Thesis, University of Vienna, Wien: 173 + 28 pp.

Jäch, M. [A.], 1982b. Neue Dryopoidea und Hydraenidae aus Ceylon, Nepal, Neu Guinea und der Türkei (Col.). Koleopterologische Rundschau 56: 89–114.

Jäch, M.A., 1983a. Podelmis xanthogramma n.sp. aus Malaya (Elmidae, Col.). Zeitschrift der Arbeitsgemeinschaft Österreichischer Entomologen 34 (3–4) [1982]: 111–112.

Jäch, M.A., 1983b. Woher stammt Ludyella corticariiformis Reitter, 1899 (Elmidae, Col.). Zeitschrift der Arbeitsgemeinschaft Österreichischer Entomologen 35 (1–2): 47–48.

Jäch, M.A., 1984a. Beitrag zur Kenntnis der Elmidae der (asiatischen) Türkei (Col.). Entomologische Blätter 80 (2–3): 136–142.

Jäch, M.A., 1984b. Chorologische und taxonomische Studien an anatolischen Krallenkäfern (Coleoptera, Elmidae). Nachrichtenblatt Bayerischer Entomologen 33 (2): 33–37.

Jäch, M.A., 1984c. Die Gattung Grouvellinus im Himalaya und in Südostasien (Elmidae, Col.). Koleopterologische Rundschau 57: 107–127.

Jäch, M.A., 1984d. Die Koleopterenfauna der Bergbäche von Südwest-Ceylon. Archiv für Hydrobiologie/Supplement 69 (2): 228–332.

Jäch, M.A., 1985. Beitrag zur Kenntnis der Elmidae und Dryopidae Neu Guineas (Coleoptera). Revue suisse de Zoologie 92 (1): 229–254.

Jäch, M.A., 1993. Microlara gen.n. mahensis sp.n., from the Seychelles (Coleoptera: Elmidae, Larainae). Zeitschrift der Arbeitsgemeinschaft Österreichischer Entomologen 45 (1–2): 15–18.

Jäch, M.A., 1994a. A taxonomic review of the Oriental species of the genus Ancyronyx Erichson, 1847 (Coleoptera, Elmidae). Revue suisse de Zoologie 101 (3): 601–622.

Jäch, M.A., 1994b. Case 2861. Elmidae Curtis, 1830 and Elmis Latreille, 1802 (Insecta, Coleoptera): proposed conservation as correct spelling and of feminine gender respectively. Bulletin of Zoological Nomenclature 51 (1): 25–27.

Jäch, M.A., 1997. Description of Laorina, a new laraine genus from Laos (Coleoptera Elmidae). Tropical Zoology 10 (2): 393–398.

Jäch, M.A., 2000. On two new Hydraenidae described by Makhan (1988). Latissimus 12: 15.

Jäch, M.A., 2002. Notes on Uralohelmis (Coleoptera: Elmidae). Entomological Problems 32 (2): 163–164.

Jäch, M.A., 2003. Ancyronyx Erichson: new faunistic records, and description of a new species from Sulawesi (Indonesia) (Coleoptera: Elmidae). Koleopterologische Rundschau 73: 255–260.

Jäch, M.A., 2004. Descriptions of two new species of Ancyronyx Erichson (Insecta: Coleoptera: Elmidae). Annalen des Naturhistorischen Museums in Wien (Ser. B) 105 [2003]: 389–395.

Jäch, M.A., 2006. Short Note. Taxonomy and nomenclature threatened by D. Makhan. Koleopterologische Rundschau 76: 360.

Jäch, M.A., 2007. Short Note. Vandalism in taxomomy. Koleopterologische Rundschau 77: 38, 60, 88, 146.

Jäch, M.A. & M. Balke, 2008. Global diversity of water beetles (Coleoptera) in freshwater, pp. 419–442. In Balian, E.V., C. Levéque, H. Segers & K. Martens (eds.), Freshwater Animal Diversity Assessment. Hydrobiologia 595: xvi + 637 pp.

Jäch, M.A. & D.S. Boukal, 1995a. Elmidae: 2. Notes on Macronychini, with descriptions of four new genera from China (Coleoptera), pp. 299–323. In Jäch, M.A. & L. Ji (eds.), Water Beetles of China. Vol. I. Zoologisch-Botanische Gesellschaft in Österreich und Wiener Coleopterologenverein, Wien: 410 pp.

Jäch, M.A. & D.S. Boukal, 1995b. Rudielmis gen.n. from South India (Coleoptera: Elmidae). Koleopterologische Rundschau 65: 149–157.

Jäch, M.A. & D.S. Boukal, 1996. Description of two new riffle beetle genera from Peninsular Malaysia (Coleoptera: Elmidae). Koleopterologische Rundschau 66: 179–189.

Jäch, M.A. & D.S. Boukal, 1997a. Description of two new genera of Macronychini: Aulacosolus and Nesonychus (Coleoptera: Elmidae). Koleopterologische Rundschau 67: 207–224.

Jäch, M.A. & D.S. Boukal, 1997b. The genus Hedyselmis Hinton (Coleoptera: Elmidae). Entomological Problems 28 (2): 111–116.

Jäch, M.A. & J. Kodada, 1996a. Graphosolus gen. nov. from Southeast Asia (Coleoptera: Elmidae). Entomological Problems 27 (2): 93–98.

Jäch, M.A. & J. Kodada, 1996b. Three new riffle beetle genera from Borneo: Homalosolus, Loxostirus and Rhopalonychus (Insecta: Coleoptera: Elmidae). Annalen des Naturhistorischen Museums in Wien (Ser. B) 98: 399–419.

Jäch, M.A. & J. Kodada, 1997. Prionosolus and Podonychus, two new genera of Macronychini (Coleoptera: Elmidae). Entomological Problems 28 (1): 9–23.

Jäch, M.A. & J. Kodada, 2006. Elmidae, pp. 60–61. In Löbl, I. & A. Smetana (eds.), Catalogue of Palaearctic Coleoptera. Vol. 3. Apollo Books, Stenstrup: 690 pp.

Jäch, M.A., J. Kodada & F. Čiampor, 2006. Elmidae, pp. 432–440. In Löbl, I. & A. Smetana (eds.), Catalogue of Palaearctic Coleoptera. Vol. 3. Apollo Books, Stenstrup: 690 pp.

Jäch, M.A. & A. Pütz, 2001. Erichia is a cephallobyrrhine limnichid! (Coleoptera: Limnichidae). Bollettino della Società entomologica italiana 133 (3): 195–197.

Jäch, M.A. & A.E.Z. Short, 2009. Notes on two remarkable water beetle descriptions recently published by Makhan (Coleoptera: Hydraenidae, Hydrophilidae). Latissimus 25: 17.

Jacquelin du Val, C., 1859. Genera des Coléoptères d'Europe comprenant leur classification en familles naturelles, la description de tous les genres, des Tableaux synoptiques destinés à faciliter l'étude, le Catalogue de toutes les espèces de nombreux dessins au trait de caractéres. Tome II. Deyrolle, Paris [1857–1859]: 287 pp., 67 pls.

Janssens, E., 1955. Mission E. Janssens & R. Tollet en Gréce (juillet–aout 1953), 11e note. Coleoptera-Dryopidæ et remarques biogéographiques sur la repartition des Coléoptères torrenticoles. Bulletin de l'Institut Royal des Sciences Naturelles de Belgique 31 (68): 1–7.

Janssens, E., 1956. Contribution à l'étude des insectes torrenticoles (Coleoptera Elmidae). Troisième note. Bulletin de l'Institut Royal des Sciences Naturelles de Belgique 32 (44): 1–4.

Janssens, E., 1957. Contribution à l'étude des insectes torrenticoles. V. – Elmidae de la Cordillère des Andes. Bulletin de l'Institut Royal des Sciences Naturelles de Belgique 33 (16): 1–8.

Janssens, E., 1959a. Contribution à l'étude de la faune d'Afghanistan. 20. Hydraenidae et Elmidae. Bulletin de l'Institut Royal des Sciences Naturelles de Belgique 35 (45): 1–8.

Janssens, E., 1959b. Une campagne hydrobiologique en Grèce (avril-mai 1957). Études sur les Coléoptères hydrobates. Bulletin de l'Institut Royal des Sciences Naturelles de Belgique 35 (21): 1–32.

Janssens, E., 1961. Contribution à l'étude de la faune de l'Afghanistan. 63. – Coléoptères torrenticoles (Hydraenidae & Elmidae). Bulletin de l'Institut Royal des Sciences Naturelles de Belgique 37 (17): 1–7.

Janssens, E., 1962. Coleoptera Elmidae. Mission zoologique de l'I.R.S.A.C. en Afrique orientale. (P. Basilewski et N. Leleup, 1957). Annales / Musée Royal de l'Afrique Centrale Tervuren, Sciences zoologiques 107: 431–434.

Jeannel, R., 1950. Un Elmide cavernicole du Congo Belge [Coleoptera Dryopidae]. Revue Française d'Entomologie 17 (3): 168–172.

Jeng, M.-L. & P.-S. Yang, 1991. Elmidae of Taiwan. Part I: Two new species of the genus Stenelmis (Coleoptera: Dryopoidea) with notes on the group of Stenelmis hisamatsui. Entomological News 102 (5): 236–252.

Jeng, M.-L. & P.-S. Yang, 1998. Taxonomic review of the genus Grouvellinus Champion (Coleoptera: Elmidae) from Taiwan and Japan. Proceedings of the Entomological Society of Washington 100 (3): 526–544.

Jung, S.W. & Y.J. Bae, 2012. Riffle beetle genus Leptelmis (Coleoptera Elmidae) in Korea with descriptions of two new species. Animal Cells and Systems 16 (3): 254–259.

Jung, S.W., M.A. Jäch & Y.J. Bae, 2015. Review of the Korean Elmidae (Coleoptera: Dryopoidea) with descriptions of three new species. Aquatic Insects 36 (2): 93–124.

Jung, S.W., Y. Kamite & Y.J. Bae, 2011. Description of Optioservus gapyeongensis new species and Heterlimnius hasegawai (Nomura) (Coleoptera: Elmidae) new to Korea. Entomological Research 41 (5): 178–184.

Kamite, Y., 2009. A revision of the genus Heterlimnius Hinton (Coleoptera, Elmidae). Japanese Journal of Systematic Entomology 15 (1): 199–226.

Kamite, Y., 2011. Three new species of the genus Heterlimnius (Coleoptera, Elmidae) from Asia. Japanese Journal of Systematic Entomology 17 (2): 409–414.

Kamite, Y., 2012. Three new species and a new combination of the genus *Heterlimnius* from Asia (Coleoptera: Elmidae). Koleopterologische Rundschau 82: 291–299.

Kamite, Y., T. Ogata & M. Satô, 2006. A new species of the genus Zaitzeviaria (Coleoptera, Elmidae) from Tsushima Islands, Japan. Japanese Journal of Systematic Entomology 12 (1): 149–153.

King, R.L., 1865. Description of Australian species of Georyssides and Parnides. The Transactions of the Entomological Society of New South Wales 1: 158–161.

Knie, J., 1975a. Ein Beitrag zur Verbreitung und Ökologie von Hydraena excisa Kiesw. Decheniana 127: 263–264.

Knie, J., 1975b. Elmis minuta n. sp. – eine neue Art des „Elmis maugetii Artenkreises" (Coleoptera: Dryopoidea). Gewässer und Abwässer 57–58: 141–146.

Knie, J., 1977. Ökologische Untersuchung der Käferfauna von ausgewählten Fließgewässern des Rheinischen Schiefergebirges (Insecta: Coleoptera). Decheniana. Verhandlungen des Naturhistorischen Vereins der Rheinlande und Westfalens 130: 151–221.

Knie, J., 1978. Untersuchungen an Elmis maugetii var. hungarica var.n. und ihre Abgrenzung gegenüber Elmis maugetii Latreille, 1798 und Elmis maugetii var. megerlei (Duftschmid, 1805) (Coleoptera: Dryopoidea). Folia Entomologica Hungarica (Rovartani Közlemények) (Series Nova) 31 (1): 61–67.

Kodada, J., 1992. Pseudamophilus davidi sp.n. from Thailand (Coleoptera: Elmidae). Linzer biologische Beiträge 24 (1): 359–356.

Kodada, J., 1993a. Dryopomorphus siamensis sp. nov., a new riffle beetle from Thailand (Coleoptera: Elmidae) and remarks on the morphology of the mouthparts and hind wing venation of D. bishopi Hinton. Entomological Problems 24 (1): 51–58.

Kodada, J., 1993b. Jaechomorphus, a new riffle beetle genus from Sumatra (Coleoptera: Elmidae: Larainae). Entomological Problems 24 (2): 1–17.

Kodada, J. & F. Čiampor, 2000. Review of the genus Vietelmis (Coleoptera: Elmidae). Entomological Problems 31 (1): 65–90.

Kodada, J. & F. Čiampor, 2003. Okalia globosa, a new genus and species of Macronychini from Malaysia (Insecta: Coleoptera: Elmidae). Revue suisse de Zoologie 110 (4): 783–795.

Kodada, J., T. Derka & F. Čiampor, 2012. Description of Jolyelmis spangleri a new species from Churí-tepui (Chimantá Massif, Venezuela), with a description of the larva of J. spangleri and J. reitmaieri (Insecta: Coleoptera: Elmidae). Zootaxa 3223: 1–23.

Kodada, J. & M.A. Jäch, 1999. Roraima carinata gen. et sp.nov. and Neblinagena doylei sp.nov., two Larainae from Mount Roraima, Venezuela (Coleoptera: Elmidae). Entomological Problems 30 (1): 13–29.

Kodada, J. & M.A. Jäch, 2005. 18.2. Elmidae Curtis, 1830, pp. 471–496. In Beutel, R.G. & R.A.B. Leschen (eds.), Handbook of Zoology. Vol. IV (38), Coleoptera, Beetles. Vol. 1: Morphology and Systematics (Archostemata, Adephaga, Myxophaga, Polyphaga partim). Walter de Gruyter, Berlin, New York: XI + 567 pp.

Kodada, J., M.A. Jäch & F. Čiampor, 2014. Ancyronyx reticulatus and A. pulcherrimus, two new riffle beetle species from Borneo, and discussion about elmid plastron structures (Coleoptera: Elmidae). Zootaxa 3760 (3): 383–395.

Kolbe, H.J., 1898. Käfer und Netzflüger Ost-Afrikas. In Möbius, K. (ed.), Deutsch-Ost-Afrika. Wissenschaftliche Forschungsresultate über Land und Leute unseres ostafrikanischen Schutzgebietes und der angrenzenden Länder. IV. Die Thierwelt Ost-Afrikas und der Nachbargebiete. Wirbellose Thiere. Dietrich Reimer (Ernst Vohsen), Berlin: 1–267 [Käfer] + 1–42 [Netzflügler], 5 pls.

Kolenati, F.A., 1846. Insecta Caucasi. Coleoptera, Dermaptera, Lepidoptera, Neuroptera, Mutillidae, Aphaniptera, Anoplura. Meletemata Entomologica 5: 1–169.

Konô, H., 1934. Die Dryopiden aus Japan. Insecta Matsumurana 8 (3): 124–128.

Konô, H., 1936. H. Sauter's Formosa-Ausbeute: Dryopidae. (Coleoptera.) Arbeiten über morphologische und taxonomische Entomologie aus Berlin-Dahlem 3 (2): 121–122.

Kundrata, R., M. Bocakova & L. Bocak, 2013. The phylogenetic position of Artematopodidae (Coleoptera: Elateroidea), with description of the first two Eurypogon species from China. Contributions to Zoology 82 (4): 199–208.

Kuwert, A., 1889. General-Uebersicht der Riolus- und Esolus-Arten des europäischen- und Mittelmeerfaunengebiets. Societas entomologica 4 (3): 24–26, 4 (4): 32–33.

Kuwert, A., 1890. Bestimmungstabelle der Parniden Europas, der Mittelmeerfauna, sowie der angrenzenden Gebiete. Verhandlungen der Zoologisch-Botanischen Gesellschaft in Wien 40: 15–54.

Lacordaire, T., 1854. Histoire des insectes. Genera des Coléoptères ou exposé méthodique et critique de tous les genres proposés jusqu'ici dans cet ordre d'insectes. Vol. 2. Librairie Encyclopédique de Roret, Paris: 548 pp.

Lafer, G.S., 1980. Zhuki-pritsepyshi (Coleoptera, Dryopidae) Dalnego Vostoka SSSR. Fauna presnykh vod Dalnego Vostoka SSSR 1980: 44–53.

Lambert, P., C.A. Maier & R.A.B. Leschen, 2015. A new species and lectotype designations for New Zealand Hydora (Elmidae: Larainae) with a world checklist. New Zealand Entomologist 38 (1): 38–57.

La Rivers, I., 1949a. A new species of Microcylloepus from Nevada (Coleoptera: Dryopidae). Entomological News 60 (8): 205–209.

La Rivers, I., 1949b. A new subspecies of Stenelmis from Nevada (Coleoptera: Dryopidæ). Proceedings of the Entomological Society of Washington 51 (5): 218–224.

Laššová, K., F. Čiampor & Z. Čiamporová-Zaťovičová, 2014. Two new Larainae species from Guayana region, Venezuela (Coleoptera: Elmidae). Zootaxa 3753 (2): 187–195.

Latreille, P.A., 1802. Histoire naturelle des Fourmis, et recueil de Mémoires et d'Observations sur les Abeilles, les Araignées, les Faucheurs, et autres insectes. Impr. Crapelet (chez T. Barois), Paris: xvi + 445 pp., 12 pls.

Latreille, P.A., 1807. Genera crustaceorum et insectorum secundum ordinem naturalem in familias disposita, iconibus exemplisque plurimis explicata. Vol. 2. A. Kœnig, Parisiis et Argentorati: 280 pp. [this book is dated 1806 by various authors, see e.g. Bouchard et al. (2011: 479), who refer to the "Journal général de la littérature de France" [1806] 9 (11), p. 321, where this book is announced, without providing clear evidence, that it was actually printed in 1806].

Latreille, P.A., 1816. Les crustacés, les arachnides et les insectes. In Cuvier, G. (ed.), Le Règne Animal distribué d'après son organisation, pour servir de base à l'histoire naturelle des animaux et d'introduction a l'anatomie comparée. Vol. 3. Deterville, Paris [1817]: xxix + 653 pp. [publication date suggested by Roux (1976) and other authors listed therein].

Lawrence, J.F. & A.F. Newton, 1982. Evolution and Classification of Beetles. Annual Review of Ecology and Systematics 13: 261–290.

Lea, A.M., 1895. Descriptions of new species of Australian Coleoptera. The Proceedings of the Linnean Society of New South Wales (Ser. 2) 9 (4): 589–634.

Leconte, J.L., 1850. IV. General remarks upon the Coleoptera of Lake Superior. In Agassiz, L. (ed.), Lake Superior: its physical character, vegetation, and animals, compared with those of other and similar regions. Gould, Kendall and Lincoln, Boston: 201–242 [+239*–242*].

Le Conte, J.L., 1852. Synopsis of the Parnidae of the United States. Proceedings of the Academy of Natural Sciences of Philadelphia 6: 41–45.

Leconte, J.L., 1861. Classification of the Coleoptera of North America. Part I. Smithsonian Institution, Washington [1861–1862]: 286 pp. [only pp. 1–208 issued in 1861].

Leconte, J.L., 1863. New species of North American Coleoptera. Part I. Smithsonian Miscellaneous Collections 6 (167): 1–86.

Leconte, J.L., 1866. Additions to the coleopterous fauna of the United States. No. 1. Proceedings of the Academy of Natural Sciences of Philadelphia 1866: 361–394.

Leconte, J.L., 1874. Descriptions of new Coleoptera chiefly from the Pacific slope of North America. Transactions of the American Entomological Society 5: 43–72.

Leconte, J.L., 1881. Elmis concolor, p. 72. In Le Conte [sic], J.L. & G.H. Horn (eds.), Description of new species of North American Coleoptera. Transactions of the Kansas Academy of Science 7: 71–74.

Leng, C.W., 1920. Catalogue of the Coleoptera of America, north of Mexico. John D. Sherman Jr., Mount Vernon, New York: XI + 470 pp.

Li, J.K., 1992. The Coleoptera fauna of northeast China. Jilin Education Publishing House, Jilin: 205 pp. [in Chinese, English title].

Lingafelter, S.W. & E.H. Nearns, 2013. Elucidating Article 45.6 of the International Code of Zoological Nomenclature: A dichotomous key for the determination of subspecific or infrasubspecific rank. Zootaxa 3709 (6): 597–600.

Linnæus, C., 1758. Systema naturæ per regna tria naturæ, secundum classes, ordines, genera, species, cum caracteribus, differentiis, synonymis, locis. Editio decima, reformata. Vol. 1. Impensis Direct. Laurentii Salvii, Holmiæ: 824 pp.

Madge, R.B. & R.D. Pope, 1980. The valid family-group name based on Elmis Latreille (Coleoptera: Dryopoidea). Entomologist's Gazette 31 (4): 255–259.

Maier, C.A., 2012. Elachistelmis gen.n. (Coleoptera: Elmidae: Elminae) from Suriname, with description of two new species. Zootaxa 3500: 61–69.

Maier, C.A., 2013. A revision of the Larainae (Coleoptera, Elmidae) of Venezuela, with description of nine new species. ZooKeys 329: 33–91.

Maier, C.A. & A.E.Z. Short, 2014. Hexanchorus bifurcatus sp. nov., a new tepui riffle beetle (Coleoptera: Elmidae: Larainae) from Tafelberg, Suriname. Zootaxa 3895 (1): 137–143.

Maier, C.A. & P. Spangler, 2011. Hypsilara royi gen.n. and sp.n. (Coleoptera, Elmidae, Larainae) from Southern Venezuela, with a revised key to Larainae of the Western Hemisphere. ZooKeys 116: 25–36.

Manzo, V., 2003. A new species of Macrelmis Motschulsky from Argentina (Coleoptera: Elmidae). Aquatic Insects 25 (3): 169–175.

Manzo, V., 2005. Key to the South America [sic] genera of Elmidae (Insecta: Coleoptera) with distributional data. Studies on Neotropical Fauna and Environment 40 (3): 201–208.

Manzo, V., 2006. A review of the American species of Xenelmis Hinton (Coleoptera: Elmidae), with a new species from Argentina. Zootaxa 1242: 53–68.

Manzo, V., 2013. Los élmidos de la región Neotropical (Coleoptera: Byrrhoidea: Elmidae): diversidad y distribución. Revista de la Sociedad Entomológica Argentina 72 (3–4): 199–212.

Manzo, V. & M. Archangelsky, 2012. Two new species of Elmidae (Coleoptera) from Argentina. Zootaxa 3478: 267–281.

Manzo, V. & N. Moya, 2010. Description of the first South American species of Neocylloepus Brown (Coleoptera: Elmidae): N. chaparensis sp. nov. from Bolivia. Studies on Neotropical Fauna and Environment 45 (3): 129–138.

Mařan, J., 1939. Noví palaearktičtí Helminthini. Neue palearktische Helminthini. Časopis České Společnosti Entomologické (Acta Societatis Entomologicae Bohemiae) 36: 40–42 [in German, introduction and second title in Czech].

Marsham, T., 1802. Entomologia Britannica, sistens insecta Britanniæ indigena, secundum methodum Linnæanam disposita. Tomus I. Coleoptera. Wilks et Taylor, Londini: xxxi + 547 + [1] pp.

Martin, J.O., 1927. A new Helmis (Coleoptera-Helmidæ) from the Northwest. The Pan-Pacific Entomologist 4 (2): 68.

Mascagni, A. & S. Calamandrei, 1992. Catalogo sistematico, geonemico e sinonimico dei Dryopoidea italiani (Insecta, Coleoptera: Psephenidae, Heteroceridae, Limnichidae, Dryopidae, Elminthidae). Redia 75 (1): 123–136.

Matsumura, S., 1915. Taxonomy of Insects. Vol. 2. Keiseisha-shoten, Tokyo: 316 + 20 + 10 pp., 5 pls. [in Japanese].

Matsumura, S., 1916. Hirata-doromushi (Betelmis Japonicus Mats.) ni tsuki. The insect world XX (1): 4–7 [in Japanese and English].

Melsheimer, F.E., 1844. Descriptions of new species of Coleoptera of the United States. Proceedings of the Academy of Natural Sciences, Philadelphia 2 (4): 98–118.

Melsheimer, F.V., 1806. A Catalogue of Insects of Pennsylvania. Part 1. W.D. Lepper, Hanover: vi + 60 pp.

Mičetić Stanković, V., M.A. Jäch & M. Kučinić, 2015. Annotated checklist of Croatian riffle beetles (Insecta: Coleoptera: Elmidae). Natura Croatica 24 (1): 93–109.

Minckley, C.O. & J.E. Deacon, 1975. Foods of the Devil's Hole pupfish, Cyprinodon diabolis (Cyprinodontidae). The Southwestern Naturalist 20 (1): 105–111.

Miranda, G.S., B.H.L. Sampaio & M.I. da S. dos Passos, 2012. Two new species of Austrolimnius Carter & Zeck (Insecta: Coleoptera: Elmidae) from Southeastern Brazil. Zootaxa 3389: 14–24.

Monte, C. & A. Mascagni, 2012. Review of the Elmidae of Ecuador with the description of ten new species (Coleoptera: Elmidae). Zootaxa 3342: 1–38.

Moog, O. & M.A. Jäch, 2003. Elmidae. In Moog, O. (ed.), Fauna Aquatica Austriaca. Edition 2002. Wasserwirtschaftskataster, Bundesministerium für Land- und Forstwirtschaft, Umwelt und Wasserwirtschaft, Wien: [5] + 2 + 12 + 10 + 10 + 10 pp.

Motschoulsky, V., 1851. Lettre a Monsieur le Dr. Renard, secrétaire de la Société Impériale des Naturalistes de Moscou. Bulletin de la Société Impériale des Naturalistes de Moscou XXIV (II): 648–657.

Motschoulsky, V., 1869. Genres et espèces d'Insectes, publiés dans différents ouvrages par Victor Motschoulsky. Horae Societatis entomologicae rossicae VI (Supplement) [1868]: 1–46 [publication date suggested by Herman (2001: 3972); pp. 47–118 issued in 1870].

Motschulsky, V. de, 1853. Hydrocanthares de la Russie. Imprimerie de la Société de Litérature Finnoise, Helsingfors: 15 pp.

Motschulsky, V. de, 1854. Lettre de M. de Motschulsky à M. Ménétries. Études Entomologiques 3: 1–15.

Motschulsky, V. de, 1860. Insectes des Indes orientales, et de contrées analogues. Études Entomologiques 8 [1859]: 25–118, pl. 1.

Müller, P.W.J., 1806a. IV. Beschreibung der um Odenbach im Departement von Donnersberg beobachteten Schlammkäfer, Limnius Illig. Magazin für Insektenkunde 5: 184–206.

Müller, P.W.J., 1806b. V. Macronychus, Krallenkäfer. Eine neue Käfergattung. Mit der Beschreibung einer neuen Art von Hakenkäfer, Parnus. Magazin für Insektenkunde 5: 207–220.

Müller, P.W.J., 1817. IV. Bemerkungen über einige Insekten. Magazin der Entomologie 2: 266–289.

Müller, P.W.J., 1821. III. Neue Insekten. Magazin der Entomologie 4: 184–230.

Mulsant, E. & C. Rey, 1872. Histoire naturelle des Coléoptères de France. Improsternés, Uncifères, Diversicornes, Spinipèdes. Deyrolle, Paris: [2] + 18 + 58 + 40 + 58 pp., 2 pls.

Musgrave, P.N., 1932. Notes on Helmidae (Coleoptera) taken in the Tennessee Great Smoky Mountains, with description of a new species. Proceedings of the Entomological Society of Washington 34 (5): 79–81.

Musgrave, P.N., 1933. New species of Helmidae (Coleoptera). Proceedings of the Entomological Society of Washington 35 (4): 54–57.

Musgrave, P.N., 1935. Two new Elmidae from Puerto Rico with description of new genus (Coleoptera). Proceedings of the Entomological Society of Washington 37 (2): 32–35.

Musgrave, P.N., 1940. A new name: Elmidae (Coleoptera). Proceedings of the Entomological Society of Washington 42 (2): 48.

Neave, S.A., 1939. Nomenclator Zoologicus. Vol. II. Richard Clay and Company Ltd., Bungay: 1025 pp.

Nikitsky, N. B., A. A. Prokin & M. I. Shapovalov, 2010. Family Elmidae, p. 128. In Zamotailov, A. S. & N. B. Nikitsky (eds.), [English title: Coleopterous insects (Insecta, Coleoptera) of the Republic of Adygheya (annotated catalogue of species). Fauna conspecta of Adygheya]. No. 1. Adyghei State University Publishers, Maykop: 404 pp. [in Russian].

Nomura, S., 1957a. Drei neue Dryopiden-Arten aus Japan. Akitu, Transactions of the Kyoto Entomological Society 6 (1): 1–5.

Nomura, S., 1957b. Mordellid- and elmid-beetles of Yakushima (Coleoptera). The Entomological Review of Japan 8 (2): 40–44.

Nomura, S., 1958a. Drei neue Stenelmis-Arten aus Japan. (Coleoptera, Elmidae). The Entomological Review of Japan 9 (2): 41–45, pl. 8.

Nomura, S., 1958b. Notes on the Japanese Dryopoidea (Coleoptera), with two species from Saghalien. Tôhô-Gakuhô 8: 45–59, 2 pls.

Nomura, S., 1959. Notes on the Japanese Dryopoidea (Coleoptera). II. Tôhô-Gakuhô 9: 33–38, 1 pl.

Nomura, S., 1960. Notes on the Japanese Dryopoidea (Coleoptera) III. Akitu, Transactions of the Kyoto Entomological Society 9: 34–36.

Nomura, S., 1961. Elmidae found in subterranean waters of Japan. Akitu, Transactions of the Kyoto Entomological Society 10 (1–2): 1–3.

Nomura, S., 1962. Some new and remarkable species of the Coleoptera from Japan and its adjacent regions. Tôhô-Gakuhô 12: 35–51, 2 pls.

Nomura, S., 1963. Notes on the Dryopoidea (Coleoptera) IV. Tôhô-Gakuhô 13: 41–56, 1 pl.

Nomura, S. & K. Baba, 1961. Two new elmid-species of Niigata prefecture, Japan (Coleoptera). Akitu, Transactions of the Kyoto Entomological Society 10 (1–2): 4–6.

Nomura, S. & T. Nakane, 1958. A new species of the genus Stenelmis from Japan (Coleoptera: Elmidae). Akitu, Transactions of the Kyoto Entomological Society 7: 81–82.

Novak, P., 1952. Kornjaši Jadranskog primorja (Coleoptera). Yugoslavian Academy of Sciences and Arts, Zagreb: 521 pp.

Ogawa, N., 2013. A new record of Paramacronychus granulates [sic] (Coleoptera, Elmidae) from Yakushima Island, Japan. Elytra (New Series) 3 (1): 65–66.

Olmi, M., 1969. Notizie ecologiche su Esolus angustatus (Ph. Müller) con considerazioni sinonimiche. Bolletino della Società Entomologica Italiana 99–101 (7–8): 232–237.

Olmi, M., 1975. Descrizione di due nuove specie italiane appartenenti al genere Esolus Mulsant et Rey (Coleoptera Elminthidae). Annali del Museo Civico di Storia Naturale Giacomo Doria 80: 232–237.

Olmi, M., 1976. Fauna d'Italia. Vol. XII, Coleoptera Dryopidae – Elminthidae. Calderini, Bologna: 280 pp.

Olmi, M., 1981. Results of the Czechoslovak-Iranian entomological expedition to Iran (Together with results of collections made in Anatolia in 1947 and 1970) Coleoptera, Dryopoidea : Dryopidae and Elminthidae. Acta entomologica Musei Nationalis Pragae 40: 337–339.

Özdikmen, H., 2005. Hintoniella, a replacement name for the preoccupied generic name Helonastes Hinton, 1968 (Coleoptera: Byrrhoidea: Elmidae). New Zealand Journal of Zoology 32 (4): 233.

Özdikmen, H., 2008. Nomenclatural changes for some preoccupied harvestman genus group names (Arachnida: Opiliones). Turkish Journal of Arachnology 1 (1): 37–43.

Pakulnicka, J. & E. Biesiadka, 2011. Water beetles (Coleoptera) of Olsztyn (Poland), pp. 305–315. In Dykiewicz, P, L. Jerzak, J. Böhner & B. Kavanagh (eds.), Urban Fauna. Studies of animal biology, ecology and conservation in European cities. University of Technology and Life Sciences in Bydgoszcz, Bydgoszcz.

Palatov, D.M., 2014. New data on the benthic invertebrate fauna in fresh waters of Kunashir Island. Vladimir Ya. Levanidov's biennial memorial meetings 6: 509–522 [in Russian].

Panzer, G.W.F., 1793. Faunae insectorum Germanicae initia oder Deutschlands Insecten. Fascicle 7. Felsecker, Nürnberg: 1–8, 8 pls.

Pascoe, F.P., 1877. XI.–Descriptions of new genera and species of New-Zealand Coleoptera.–Part IV. The Annals and Magazine of Natural History (Ser. 4) xix (110): 140–147.

Passos, M.I.S. & M. Felix, 2004a. A new species of Macrelmis Motschulsky from southeastern Brazil (Coleoptera: Elmidae: Elminae). Studies on Neotropical Fauna and Environment 39 (1): 49–51.

Passos, M.I.S. & M. Felix, 2004b. Description of a new species of Cylloepus Erichson from southeastern Brazil (Coleoptera, Elmidae). Revista Brasileira de Entomologia 48 (2): 181–183.

Passos, M.I.S. dos, A.S. Fernandes, N. Hamada & J.L. Nessimian, 2010a. Insecta, Coleoptera, Elmidae, Amazon region. Check List 6 (4): 538–545.

Passos, M.I.S. dos, B.H.L. Sampaio, J.L. Nessimian & N. Ferreira, 2010b. Elmidae (Insecta: Coleoptera) do Estado do Rio de Janeiro: lista de espécies e novos registros. Arquivos do Museu Nacional, Rio de Janeiro 67 (3–4) [2009]: 377–382 [correct publication date according to N. Ferreira, pers. comm.].

Paulian, R., 1959. Recherches sur les insectes d'importance biologique a Madagascar (XXIX à XXXIII) [XXXI Un Helmidae cavernicole malgache (Coléoptère)]. Mémoires de l'Institut scientifique de Madagascar (Ser. E) 11: 1–16.

Peck, S.B., J. Cook & J.D. Hardy, 2003. Beetle fauna of the island of Tobago, Trinidad and Tobago, West Indies. Insecta Mundi 16 (1–3) [2002]: 9–23.

Perez Arcas, L., 1865. Insectos nuevos ó poco conocidos de la Fauna española. Segunda parte. Revista de los Progresos de las Ciencias Exactas, Físicas y Naturales 15 (7): 413–444.

Peris, D.C., C.A. Maier & A. Sánchez-García, 2015. Elmadulescens, p. 283, Elmadulescens rugosus, p. 283. In Peris, D.C., C.A. Maier, A. Sánchez-García, & X. Delclòs, The oldest known riffle beetle (Coleoptera: Elmidae) from Early Cretaceous Spanish amber. Comptes Rendus Palevol 14: 181–186.

Perkins, P.D. & W.E. Steiner, 1981. Two new Peruvian species of the riffle beetle genus Xenelmis (Coleoptera: Elmidae). The Pan-Pacific Entomologist 57 (1): 306–312.

Peyerimhoff, P. de, 1929. Nouveaux Coléoptères du Nord-Africain, Soixante-dixième note. Faune du Hoggar et des massifs voisins (suite). Bulletin de la Société Entomologique de France 1929 (10): 168–172.

Philippi, R.A., 1864. Ein Käferchen, das als Gewürz dient. Entomologische Zeitung [Stettin] 25 (1–3): 93–96.

Pic, M., 1894. Note sur les Elmides. Revue d'Entomologie 13: 193–195.

Pic, M., 1895. Notes sur des Coléoptères rares ou nouveaux d'Algérie. Annales de la Société Entomologique de France 64, Bulletin des Séances et Bulletin bibliographique de la Société Entomologique de France 1895 (4): CXXVI–CXXX.

Pic, M., 1898. Coléoptères rares ou nouveaux récoltés par M. Maurice Pic cette année dans les Alpes. [Bulletin de la] Société d'histoire naturelle d'Autun 11 (2): 154–156.

Pic, M., 1900a. Description d'un nouveau genre d'Elmides, de Tunisie [Col.]. Bulletin de la Société Entomologique de France 1900 (13): 266–267.

Pic, M., 1900b. Note sur le genre «Esolus» Muls-Rey [sic]. L'Échange, Revue Linnéenne 16 (188): 60.

Pic, M., 1901. Notes diverses et diagnoses. L'Échange, Revue Linnéenne 17 (193): 2–4.

Pic, M., 1905. Diagnoses de Coléoptères algériens. L'Échange, Revue Linnéenne 21 (247): 145–148.

Pic, M., 1923. Nouveautés diverses. Melanges exotico-entomologiques 39: 1–32.

Pic, M., 1930. Spedizione di S.A.R. il Duca degli Abruzzi alle sorgenti dell'Uebi Scebeli. – Risultati zoologici. Un Helmidae africain nouveau. Annali del Museo Civico di Storia Naturale Giacomo Doria 55: 23–24.

Pic, M., 1939. Coléoptères nouveaux d'Egypte et du Sinaï. Bulletin de la Société Fouad 1er d'Entomologie 23: 143–149.

Pic, M., 1950. Deux nouveaux Coléoptères d'Egypte et Nubie. Bulletin de la Société Fouad 1er d'Entomologie 34: 23–24.

Plachý, J., 2006. Systematická revízia vietnamských a juhočínskych druhov rodu Leptelmis Sharp, 1888 (Insecta: Coleoptera: Elmidae). Thesis, Comenius University, Bratislava: 41 pp.

Poole, R.W. & P. Gentili, 1996. Nomina Insecta Nearctica: A Check List of the Insects of North America. Vol. 1: Coleoptera and Strepsiptera. Entomological Information Services, Rockville (Maryland): 827 pp.

Przewoźny, M., P. Buczyński, C. Greń, R. Ruta & G. Tończyk, 2011. New localities of Elmidae (Coleoptera: Byrrhoidea), with a revised checklist of species occurring in Poland. Polish Journal of Entomology (Polskie Pismo Entomologiczne) 80 (2): 365–390.

Przewoźny, M. & A.S. Fernandes, 2012. Portelmis guianensis sp. nov. from French Guiana (Coleoptera: Elmidae). Zootaxa 3196: 58–63.

Rafinesque, C.S., 1815. Analyse de la nature ou tableau de l'univers et des corps organisés. C.S. Rafinesque, Palerme: 224 pp.

Reiche, L., 1879. Descriptions de quelques nouvelles espèces de Géorissides, Parnides et Hétérocérides propres à la faune européenne. Annales de la Société Entomologique de France (Ser. 5) 9: 237–239.

Reitter, E., 1883. Coleopterologische Notizen. Deutsche Entomologische Zeitschrift 27 (1): 74–75.

Reitter, E., 1885. Neue Coleopteren aus Europa und den angrenzenden Ländern, mit Bemerkungen über bekannte Arten. Deutsche Entomologische Zeitschrift 29 (2): 353–392.

Reitter, E., 1886. Drei neue Elmiden von Sumatra. Notes from the Leyden Museum 8: 213–214.

Reitter, E., 1887. Neue Coleopteren aus Europa, den angrenzenden Ländern und Sibirien, mit Bemerkungen über bekannte Arten. Dritter Theil. Deutsche Entomologische Zeitschrift 31 (1): 241–288.

Reitter, E., 1889. Neue Coleopteren aus dem Leydener Museum. Notes from the Leyden Museum 9: 3–9.

Reitter, E., 1895. Beschreibung neuer oder wenig gekannter Coleopteren aus der Umgebung von Akbes in Syrien. Wiener Entomologische Zeitung XIV (III): 79–88.

Reitter, E., 1899. Abbildungen und Beschreibungen neuer oder wenig gekannter Coleopteren aus der palaearctischen Fauna. Wiener Entomologische Zeitung 18 (9): 282–287, pl. IV.

Reitter, E., 1901. Coleopterologische Notizen. Wiener Entomologische Zeitung XX (III): 57–59.

Reitter, E., 1906. Dreizehn neue Coleopteren aus der palaearktischen Fauna. Wiener Entomologische Zeitung 25 (8–9): 237–244.

Reitter, E., 1907. Übersicht der mir bekannten Stenelmis-Arten aus der paläarktischen Fauna. (Col.). Deutsche Entomologische Zeitschrift 1907 (5): 483–484.

Reitter, E., 1910. Elmis zoufali n. sp. Wiener Entomologische Zeitung 29: 36.

Rey, C., 1889. Remarques en passant. L'Échange, Revue Linnéenne 5 (57): 66–67.

Rosenhauer, W.G., 1856. Die Thiere Andalusiens nach dem Resultate einer Reise zusammengestellt, nebst den Beschreibungen von 249 neuen oder bis jetzt noch unbeschriebenen Gattungen und Arten.T. Blaesing, Erlangen: viii + 429 pp., 3 pls.

Roubal, J., 1940. Uralohelmis, genus novum helminarum Europae. Sborník Entomologického Oddělení při Zoologických Sbírkách Národního Musea v Praze (Acta Entomologica Musaei Nationalis Pragae) 18: 151–154.

Roux, C., 1976. On the dating of the first edition of Cuvier's Règne Animal. Journal of the Society for the Bibliography of Natural History 8 (1): 31.

Sainte-Claire Deville, J., 1905. Contributions a la faune française. L'Abeille, Journal d'Entomologie 30: 237–248.

Sainte-Claire Deville, J., 1919. Description d'un Limnius nouveau de France. Bulletin de la Société Entomologique de France 1919 (15): 263–264.

Samouelle, G., 1819. The entomologist's useful compendium; or an introduction to the knowledge of British insects, comprising the best means of obtaining and preserving them, and a description of the apparatus generally used; together with the genera of Linné, and the modern method of arranging the classes Crustacea, Myriapoda, spiders, mites and insects, from their affinities and structure, according to the views of Dr. Leach. Also an explanation of the terms used in entomology; a calendar of the times of appearance and usual situations of near 3,000 species of British insects; with instructions for collecting and fitting up objects for the microscope. Thomas Boys, London: 496 pp., 12 pls.

Sampaio, B.H.L., M.I. da S. dos Passos & N. Ferreira, 2011. Three new species of Cylloepus Erichson (Insecta: Coleoptera: Elmidae) from Southeastern Brazil. Zootaxa 2797: 57–64.

Sampaio, B.H.L., M.I. da S. dos Passos & N. Ferreira, 2012. Two new species of Macrelmis Motschulsky (Coleoptera: Elmidae) and a new record of Macrelmis isis [sic] (Hinton) from Southeastern Brazil. Zootaxa 3478: 164–168.

Sanderson, M.W., 1938a. A monographic revision of the North American species of Stenelmis (Dryopidae: Coleoptera). The University of Kansas Science Bulletin 25 (22): 635–717.

Sanderson, M.W., 1938b. Elmis columbiensis Angell a synonym of Zaitzevia parvulus (Horn). Journal of the Kansas Entomological Society 11 (4): 146.

Sanderson, M.W., 1953a. A revision of the Nearctic genera of Elmidae (Coleoptera). Journal of the Kansas Entomological Society 26 (4): 148–163.

Sanderson, M.W., 1953b. New species and a new genus of New World Elmidae with supplemental keys. The Coleopterists' Bulletin VII (5): 33–40.

Sanderson, M.W., 1954. A revision of the Nearctic genera of Elmidae (Coleoptera). Journal of the Kansas Entomological Society 27 (1): 1–13.

Satô, M., 1960. Aquatic Coleoptera from Amami-Ôshima of the Ryukyu Islands (I). Kontyû 28 (4): 251–254.

Satô, M., 1963a. New forms in the genus Zaitzevia Champion from the Ryukyus (Col. Elmidae). New Entomologist 12 (7): 39–41.

Satô, M., 1963b. Some aquatic beetles unrecorded from Shikoku, Japan, II (Coleoptera). Transactions of the Shikoku Entomological Society 7: 132.

Satô, M., 1964a. Description of a new elmid-beetle from the Ryukyus. Bulletin of the Japan Entomological Academy 1 (2): 11–12, pl. 3.

Satô, M., 1964b. Descriptions of new dryopoid-beetles from the Ryukyus. Bulletin of the Japan Entomological Academy 1 (7): 31–37.

Satô, M., 1973. Notes on dryopoid beetles from New Guinea. Pacific Insects 15 (3–4): 463–471.

Satô, M., 1976. Taxonomic and zoogeographical notes on the genus Zaitzeviaria and its allied genera (Coleoptera; Elminthidae). Data of reading for XV. International Congress of Entomology, Washington D.C., [August] 19–27, 1976: 4 unnumbered pp. (unpublished).

Satô, M., 1977a. Family Elminthidae, pp. 1–6. In Check-list of Coleoptera of Japan 9. The Coleopterist's Association of Japan, Tokyo.

Satô, M., 1977b. Ergebnisse der Bhutan-Expedition 1972 des Naturhistorischen Museums in Basel. Coleoptera: Fam. Hydrophilidae, Dryopidae and Elminthidae. Entomologica Basiliensia 2: 197–204.

Satô, M., 1977c. Studies on the genus Zaitzeviaria (Coleoptera, Elminthidae), I. Description of new species. Annotationes zoologicae Japonenses 50 (3): 191–194.

Satô, M., 1978. Stenelmis Dufour species of Korea (Coleoptera, Elminthidae). Annales historico-naturales Musei nationalis hungarici 70: 147–149.

Satô, M., 1999. A new Stenelmis (Coleoptera, Elmidae) from the Ryukyu Islands. The Entomological Review of Japan 54 (2): 121–123.

Satô, M., 2002. Records on Elmidae (Coleoptera) from Sri Lanka. The Entomological Review of Japan 57 (2): 165–168.

Satô, M. & T. Kishimoto, 2001. A new cavernicolous elmid (Coleoptera, Elmidae) discovered in South China. Elytra 29 (1): 75–85.

Say, T., 1824. Order Coleoptera, pp. 268–297. In Keating, W.H. (ed.), Narrative of an expedition to the source of St. Peter's River, Lake Winnepeek, Lake of the Woods, &c. &c. performed in the year 1823, by the order of the Hon. J. C. Calhoun, secretary of war, under the command of Stephen H. Long, Major U. S. T. E. Vol. II. H.C. Carey & I. Lea, Philadelphia: VI + 459 pp., pls. 1–14.

Say, T., 1825. Descriptions of new species of coleopterous insects inhabiting the United States. Journal of the Academy of Natural Sciences of Philadelphia 5 (1): 160–204.

Schaeffer, C., 1911. New Coleoptera and miscellaneous notes. Journal of the New York Entomological Society 19 (2): 113–126.

Schmude, K.L., 1992. Revision of the riffle beetle genus Stenelmis (Coleoptera: Elmidae) in North America, with notes on bionomics. Thesis, University of Wisconsin, Madison: 286 + 31 pp.

Schmude, K.L., 1999. Riffle beetles in the genus Stenelmis (Coleoptera: Elmidae) from warm springs in southern Nevada: new species, new status, and a key. Entomological News 110 (1): 1–12.

Schmude, K.L. & C.B. Barr, 1992. Stenelmis xylonastis, pp. 587–594. In Schmude, K.L., C.B. Barr & H.P. Brown (eds.), Stenelmis lignicola and Stenelmis xylonastis, two new North American species of wood-inhabiting riffle beetles (Coleoptera: Elmidae). Proceedings of the Entomological Society of Washington 94 (4): 580–594.

Schmude, K.L. & H.P. Brown, 1991. A new species of Stenelmis (Coleotpera: Elmidae) found west of the Mississippi river. Proceedings of the Entomological Society of Washington 93 (1): 51–61.

Schmude, K.L. & H.P. Brown, 1992. Stenelmis lignicola, pp. 583–587. In Schmude, K.L., C.B. Barr & H.P. Brown (eds.), Stenelmis lignicola and Stenelmis xylonastis, two new North American species of wood-inhabiting riffle beetles (Coleoptera: Elmidae). Proceedings of the Entomological Society of Washington 94 (4): 580–594.

Schöll, F., 2002. Das Makrozoobenthos des Rheins 2000. 68. Plenarsitzung – 2./3. Juli 2002 – Luxemburg (Bericht Nr. 128-d). Internationale Kommission zum Schutz des Rheins, Koblenz: 49 pp.

Schönfeldt, H. von, 1888. Stenelmis foveicollis, ein neuer japanischer Käfer. Entomologische Nachrichten 14 (13): 193–194.

Segura, M.O., M.I. da S. dos Passos, A.A. Fonseca-Gessner & C.G. Froehlich, 2013. Elmidae Curtis, 1830 (Coleoptera, Polyphaga, Byrrhoidea) of the Neotropical region. Zootaxa 3731 (1): 1–57.

Sharp, D., 1872. Descripciones de algunas especies nuevas de Coleópteros. Anales de la Sociedad Española de Historia Natural 1: 259–271.

Sharp, D., 1882. Fam. Parnidae, pp. 119–140, 1 pl. In Godman, F.D. & O. Salvin (eds.), Biologia Centrali-Americana, Insecta, Coleoptera 1 (2). London [1882–1887]: xvi + 824 pp., pls. 1–19.

Sharp, D., 1887. Supplement, pp. 774–775. In Godman, F.D. & O. Salvin (eds.), Biologia Centrali-Americana, Insecta, Coleoptera 1 (2). London [1882–1887]: xvi + 824 pp., pls. 1–19.

Sharp, D., 1888. Descriptions of some new Coleoptera from Japan. The Annals and Magazine of Natural History (Ser. 6) II (9): 242–245.

Shepard, W.D., 1990. Microcylloepus formicoideus (Coleoptera: Elmidae), a new riffle beetle from Death Valley National Monument, California. Entomological News 101 (3): 147–153.

Shepard, W.D., 1993. An annotated checklist of the aquatic and semiaquatic dryopoid Coleoptera of California. The Pan-Pacific Entomologist 69 (1): 1–11.

Shepard, W.D., 1998. Elmidae: II. Description of Orientelmis gen.n. and new synonymy in Cleptelmis Sanderson (Coleoptera), pp. 289–295. In Jäch, M.A. & L. Ji (eds.), Water Beetles of China. Vol. II. Zoologisch-Botanische Gesellschaft und Wiener Coleopterologenverein, Wien: 371 pp.

Shepard, W.D., 2004. Lotic regions of Belize and their aquatic byrrhoid Coleoptera (Dryopidae, Elmidae, Lutrochidae, Psephenidae, Ptilodactylidae). The Pan-Pacific Entomologist 80 (1–4): 53–59.

Sherborn, C.D. & F.J. Griffin, 1934. IX.–On the dates of publication of the natural history portions of Alcide d'Orbigny's 'Voyage Amérique méridionale'. The Annals and Magazine of Natural History (Ser. 10) XIII (73): 130–134.

Smith, P.B., 1989. A description of the larva of *Rhyncholimnochares kittatinniana* Habeeb (Hydrachnidia: Limnocharidae). The Canadian Entomologist 121: 445–452.

Sondermann, W., 2012. Is the elmid fauna of Colombia strongly marked by Nearctic elements? A remote analysis of genus names provided in 30 recently published benthic macroinvertebrate assessments: (Coleoptera: Byrrhoidea: Elmidae). Dugesiana 20 (2): 251–260.

Spangler, P.J., 1966. XIII – Aquatic Coleoptera (Dytiscidae; Noteridae; Gyrinidae; Hydrophilidae; Dascillidae; Helodidae; Psephenidae; Elmidae), pp. 377–443. In Patrick, R. et al. (eds.), The Catherwood Foundation Peruvian-Amazon Expedition. Monographs of the Academy of Natural Sciences of Philadelphia 14: 495 pp.

Spangler, P.J., 1980a. A new species of the riffle beetle genus Portelmis from Ecuador (Coleoptera: Elmidae). Proceedings of the Entomological Society of Washington 82 (1): 63–68.

Spangler, P.J., 1980b. V. Aquatic Coleoptera, pp. 199–213. In Roback, S.S., L. Berner, O.S. Flint, N. Nieser & P.J. Spangler (eds.), Results of the Catherwood Bolivian-Peruvian Altiplano Expedition. Part I. Aquatic insects except Diptera. Proceedings of the Academy of Natural Sciences of Philadelphia 132: 176–217.

Spangler, P.J., 1981a. Amsterdam expeditions to the West Indian Islands, Report 15. Two new genera of phreatic elmid beetles from Haiti; one eyeless and one with reduced eyes (Coleoptera, Elmidae). Bijdragen tot de Dierkunde (Contributions to Zoology) 51 (2): 375–387.

Spangler, P.J., 1981b. Pagelmis amazonica, a new genus and species of water beetle from Ecuador (Coleoptera: Elmidae). The Pan-Pacific Entomologist 57 (1): 286–294.

Spangler, P.J., 1985a. A new genus and species of riffle beetle, Neblinagena prima, from the Venezuelan tepui, Cerro de la Neblina (Coleoptera, Elmidae, Larinae). Proceedings of the Entomological Society of Washington 87 (3): 538–544.

Spangler, P.J., 1985b. A new species of the aquatic beetle genus Dryopomorphus from Borneo (Coleoptera: Elmidae: Larinae). Proceedings of the Biological Society of Washington 98 (2): 416–421.

Spangler, P.J., 1986. The status of the riffle beetle genus Lara and homonymy of the subfamily group name Larinae (Coleoptera: Elmidae). Entomological News 97 (2): 77–79.

Spangler, P.J., 1987. Case 2581. Laridae Vigors, 1825 (Aves) and Larini LeConte, 1861 (Insecta, Coleoptera): proposal to remove the homonymy. Bulletin of Zoological Nomenclature 44 (1): 25–26.

Spangler, P.J., 1990. A revision of the Neotropical aquatic beetle genus Stegoelmis (Coleoptera: Elmidae). Smithsonian Contributions to Zoology 502: iv + 52 pp.

Spangler, P.J., 1996a. A new genus and species of aquatic beetle, Caenelmis octomeria, from Kenya, Africa (Coleoptera: Elmidae: Elminae). Insecta Mundi 10 (1–4): 19–23.

Spangler, P.J., 1996b. Four new stygobiontic beetles (Coleoptera: Dytiscidae; Noteridae; Elmidae). Insecta Mundi 10 (1–4): 241–159.

Spangler, P.J., 1997. Two new species of the aquatic beetle genus Macrelmis Motschulsky from Venezuela (Coleoptera: Elmidae: Elminae). Insecta Mundi 11 (1): 1–8.

Spangler, P.J. & H.P. Brown, 1981. The discovery of Hydora, a hitherto Australian-New Zealand genus of riffle beetles, in austral South America (Coleoptera: Elmidae). Proceedings of the Entomological Society of Washington 83 (4): 596–606.

Spangler, P.J. & R.A. Faitoute, 1991. A new genus and species of Neotropical water beetle, Jolyelmis auyana, from a Venezuelan tepui (Coleoptera: Elmidae). Proceedings of the Biological Society of Washington 104 (2): 322–327.

Spangler, P.J. & P.D. Perkins, 1989. A revision of the Neotropical aquatic beetle genus Stenhelmoides (Coleoptera: Elmidae). Smithsonian Contributions to Zoology 479: iii + 63 pp.

Spangler, P.J. & S. Santiago, 1982. A new species of aquatic beetle, Disersus uncus, from Costa Rica (Coleoptera: Elmidae: Larainae), pp. 17–20. In Satô, M., Y. Hori, Y. Arita & T. Okadome (eds.), Special Issue to the Memory of Retirement of Emeritus Professor Michio Chûjô. The Association of the Memorial Issue of Emeritus Professor M. Chûjô, c/o Biological Laboratory, Nagoya Women's University, Nagoya: 185 pp.

Spangler, P.J. & S. Santiago, 1987. A revision of the Neotropical aquatic beetle genera Disersus, Pseudodisersus, and Potamophilops (Coleoptera: Elmidae). Smithsonian Contributions to Zoology 446: iv + 40 pp.

Spangler, P.J. & S. Santiago, 1991. A new species and new records from Colombia of the water beetle genus Onychelmis Hinton (Coleoptera: Elmidae: Elminae). Proceedings of the Entomological Society of Washington 93 (2): 495–498.

Spangler, P.J. & S. Santiago F. [Fragoso], 1986. Una nueva especie de Coleoptero acuatico, del género Macrelmis Motschulsky de México y Centroamerica (Coleoptera: Elmidae). Anales del Instituto de Biología de la Universidad Nacional Autónoma de México, Serie Zoología 56 (1) [1985]: 155–158.

Spangler, P.J. & S. Santiago-Fragoso, 1992. The aquatic beetle subfamily Larainae (Coleoptera: Elmidae) in Mexico, Central America and the West Indies. Smithsonian Contributions to Zoology 528: vi + 74 pp., 1 pl.

Spangler, P.J. & C.L. Staines, 2004a. Luchoelmis, a new genus of Elmidae (Coleoptera) from Chile and Argentina. Insecta Mundi 16 (4) [2002]: 215–220.

Spangler, P.J. & C.L. Staines, 2004b. Three new species of Hexanchorus Sharp, 1882 (Coleoptera: Elmidae: Larainae) from South America. Insecta Mundi 17 (1–2) [2003]: 45–48.

Statz, G., 1939. Geradflügler und Wasserkäfer der oligocänen Ablagerungen von Rott. Decheniana 99 A: 1–102, pls. I–XX.

Steffan, A.W., 1958. Die deutschen Arten der Gattungen Elmis, Esolus, Oulimnius, Riolus, Aptyktophallus (Coleoptera: Dryopoidea). Genitalmorphologisch-taxionomische Studie an Dryopoidea I. Beiträge zur Entomologie 8 (1–2): 122–178.

Steffan, A.W., 1961. Vergleichend-mikromorphologische Genitaluntersuchungen zur Klärung der phylogenetischen Verwandtschaftsverhältnisse der mitteleuropäischen

Dryopoidea (Coleoptera). Zoologische Jahrbücher, Abteilung für Systematik, Ökologie und Geographie der Tiere 88 (3): 255–354.

Stephens, J.F., 1828. Illustrations of British Entomology; or, a synopsis of indigenous insects: containing their generic and specific distinctions; with an account of their metamorphoses, times of appearance, localities, food and economy, as far as practicable. Mandibulata. Vol. II. Baldwin and Cradock, London [1828–1829]: 200 pp., pls. X–XV [only pp. 1–112 and pls. X–XII issued in 1828].

Stewart, K.W., G.P. Friday & R.E. Rhame, 1973. Food habits of hellgrammite larvae, Corydalus cornutus (Megaloptera: Corydalidae), in the Brazos River, Texas. Annals of the Entomological Society of America 66: 959–963.

Sturm, J., 1826. Catalog meiner Insecten – Sammlung. Sturm, Nürnberg: VIII + 207 + [1] + 16 + [2] pp., pls. I–IV.

Sturm, J., 1843. Catalog der Kæfer - Sammlung von Jacob Sturm. Sturm, Nürnberg: VI + 386 pp., pls. I–VI.

Sturm, J.H.C.F., 1857. Dr. Jacob Sturm's Deutschlands Fauna in Abbildungen nach der Natur mit Beschreibungen (V) XXIII. Sturm, Nürnberg: 123 pp., pls. CCCCIX–CCCCXXIV [the authorship of this volume is often erroneously credited to Jacob Sturm (1771–1848), however, volumes XX–XXIII were published by Jacob Sturm's son Johann Heinrich Christian Friedrich Sturm (1805–1862)].

Tamutis, V., B. Tamutė & R. Ferenca, 2011. A catalogue of Lithuanian beetles (Insecta, Coleoptera). ZooKeys 121: 1–494.

Telnov, D., 2004. Check-list of Latvian Beetles (Insecta: Coleoptera). In Telnov, D. (ed.), Compendium of Latvian Coleoptera. Vol. I. Entomological Society of Latvia, Rīga: 113 pp.

Theobald, N., 1937. Les insectes fossiles des terrains oligocènes de France. Bulletin Mensuel de la Société des Sciences de Nancy (Mémoires de la Société des Sciences de Nancy) (Nouvelle Série), 2: 1–473, pls. I–XXIX.

Touaylia, S., M. Bejaoui, M. Boumaiza & J. Garrido, 2010. Contribution à l'étude des Coléoptères aquatiques de Tunisie: Les Elmidae Curtis, 1830 et les Dryopidae Billberg, 1820 (Coleoptera). Nouvelle Revue d'Entomologie (Nouvelle Série) 26 (2) [2009]: 167–175.

Van Dyke, E.C., 1949. New species of North American Coleoptera. The Pan-Pacific Entomologist 25 (2): 49–56.

Victor, T. [Motschulsky, V. de], 1839. Coléoptères du Caucase et des provinces Transcaucasiennes. (Continuation). Bulletin de la Société Impériale des Naturalistes de Moscou 12 (1): 68–93, pls. V–VI.

Vigors, N.A., 1825. XXII. Observations on the natural affinities that connect the orders and families of birds. Transactions of the Linnean Society of London 14 (3): 395–517.

Villa, A. & J.B. Villa, 1833. Coleoptera Europae dupleta in collectione Villa quae pro mutua commutatione offerri possunt. Villa & Villa, Mediolani, 36 pp.

Villa, A. & J.B. Villa, 1835. Supplementum Coleopterorum Europae dupletorum. Villa & Villa, Mediolani, pp. 37–50.

Vondel, B.J. van, 2003. Haliplidae: I. Three new synonymies, pp. 285–287. In Jäch, M.A. & L. Ji (eds.), Water Beetles of China. Vol. III. Zoologisch-Botanische Gesellschaft und Wiener Coleopterologenverein, Wien: VI + 572 pp.

Waterhouse, C.O., 1876. On various new genera and species of Coleoptera. The Transactions of the Entomological Society of London 1876 (1): 11–25.

Waterhouse, C.O., 1879. New species of Cleridae and other Coleoptera from Madagascar. Cistula Entomologica 2 (22) [1875–1882]: 529–535.

Waterhouse, F.H., 1879. Descriptions of new Coleoptera of geographical interest, collected by Charles Darwin, Esq. The Journal of The Linnean Society (Zoology) 14 (78): 530–534.

White, D.S., 1978. A revision of the nearctic Optioservus (Coleoptera: Elmidae), with descriptions of new species. Systematic Entomology 3: 59–74.

White, D.S., 1982. Stenelmis morsei, a new species of riffle beetle (Coleoptera: Dryopoidea: Elmidae) from South Carolina. The Coleopterists Bulletin 36 (2): 170–174.

White, D.S. & H.P. Brown, 1976. A new species of Stenelmis from North Carolina (Coleoptera: Elmidae). The Coleopterists Bulletin 30 (2): 189–192.

Więźlak, W.W., 1987a. New species of genus Austrelmis Brown from Peru (Coleoptera, Limniidae). Polskie pismo entomologiczne (= Bulletin entomologique de Pologne) 57: 299–303.

Więźlak, W.W., 1987b. Contribution to the knowledge of African Larinae (Coleoptera, Limniidae). Polskie pismo entomologiczne (= Bulletin entomologique de Pologne) 57: 441–451.

Więźlak, W.W., 1987c. Potamodytes lokis sp.n. from Tanzania (Coleoptera, Limniidae). Annales historico-naturales Musei nationalis hungarici, Pars Zoologica 79: 119–120.

Winkler, A., 1926. Catalogus Coleopterorum regionis palaearcticae. Winkler, Wien [1924–1932]: [1] + VII + [1] + 1648 columns [2 per page] + [1] + pp. 1650–1698 [only columns 625–752 issued in 1926].

Yang, J. & Z. Zhang, 1995. S.[tenelmis] euronotata, S.[tenelmis] grossimarginata, S.[tenelmis] montana, S.[tenelmis] sinica, p. 102, S.[tenelmis] indepresa [sic], p. 103, S.[tenelmis] huangkengana Yang et Zhang, p. 105, S.[tenelmis] heteromorpha Yang et Zhang, p. 106. In Zhang, Z. & J. Yang (eds.), Coleoptera: Dryopoidea. In Zhu, T. (ed.), [English title: Insects and Macrofungi of Gutianshan, Zhejiang]. Zhejiang Science and Technology Publishing House, Hangzhou: 102–110 [in Chinese, with English summary].

Yang, J. & Z. Zhang, 2002. Elmididae, pp. 811–824. In Huang, B. (ed.), [English title: Fauna of Insects in Fujian Province of China]. Vol. 6. Fujian Scientific Technology Publishing, Fuzhou: 894 pp. [in Chinese, with English summary].

Yoshitomi, H. & M.-L. Jeng, 2013. A new species of the genus Dryopomorphus Hinton (Coleoptera, Elmidae, Larainae) from Laos. Elytra (New Series) 3 (1): 45–51.

Yoshitomi, H. & J. Nakajima, 2007. A new species of the genus Sinonychus (Coleoptera, Elmidae) from Japan. Elytra 35 (1): 96–101.

Yoshitomi, H. & J. Nakajima, 2012. A new species of the genus Sinonychus (Coleoptera, Elmidae) from Kyushu, Japan. Elytra (New Series) 2 (1): 53–60.

Yoshitomi, H. & M. Satô, 2005. A revision of the Japanese species of the genus Dryopomorphus (Coleoptera, Elmidae). Elytra 33 (2): 455–473.

Zaitzev, P., 1908. Catalogue des Coléoptères aquatiques des familles des Dryopidae, Georyssidae, Cyathoceridae, Heteroceridae et Hydrophilidae. Trudy Russkago entomologicheskago obshchestva [Horae Societatis entomologicae rossicae] 38: 283–420.

Zaitzev, P., 1910. Pars 17. Dryopidae, Cyathoceridae, Georyssidae, Heteroceridae. In Schenkling, S. (ed.), Coleopterorum Catalogus. W. Junk, Berlin: 68 pp.

Zaitzev, P.A., 1947. Vodyanye zhuki basseyna reki Zangi i nekotorykh drugikh vodoemov Armanskoy SSR. [German title: Ueber die Wasserkäfer des Zangaflusses und einiger anderen Gewässer Armeniens]. Trudy Sevanskoy Gidrobiologicheskoy Stantsii 8: 87–95 [in Russian, with German summary].

Zaitzev, P.A., 1951. Vodnye zhuky Turkmenistana [Water beetles of Turkmenistan]. Trudi Murgabskoy Gidrobiologicheskoy Stantsii I: 53–76 [in Russian].

Zaragoza Caballero, S., 1982. Una nueva subespecie de Hexanchorus gracilipes Sharp 1882 (Coleoptera: Elmidae; Larini) de Soteapa, Veracruz, México. Anales del Instituto de Biología de la Universidad Nacional Autónoma de México, Serie Zoología 52 (1) [1981]: 353–360.

Zeck, E.H., 1948. Two new species of Australian Dryopoidea (Coleoptera). The Australian Zoologist 11 (3): 277–279.

Zhang, Z., 1994. [English title: A taxonomic study on 3 genera of Chinses [sic] Elminae (Coleoptera: Elmidae)]. Thesis, Department of Plant Protection, Beijing Agricultural University: 73 + [2] pp., 36 pls. [in Chinese, with English abstract].

Zhang, Z. & W. Ding, 1995. [English title: Two new species and a new subspecies of Elmidae (Coleoptera: Dryopoidea) from China]. Entomotaxonomia 17 (Supplement): 15–19 [in Chinese, with English summary].

Zhang, Z., H. Su & C. Yang, 2003a. [English title: Four new species of Stenelmis (Coleoptera: Dryopoidea, Elmididae) from China]. Journal of China Agricultural University 8 (1): 106–108 [in Chinese, with English summary].

Zhang, Z., H. Su & C. Yang, 2003b. [English title: Three new species and one new record of Leptelmis (Coleoptera: Dryopoidea: Elmididae)]. Entomotaxonomia 25 (3): 189–194 [in Chinese, with English summary].

Zhang, Z. & J.[= C.] Yang, 1995. Coleoptera: Dryopoidea. In Zhu, T. (ed.), [English title: Insects and Macrofungi of Gutianshan, Zhejiang]. Zhejiang Science and Technology Publishing House, Hangzhou: 102–110 [in Chinese, with English summary].

Zhang, Z. & C. Yang, 2003. [English title: Four new species of Stenelmis of Elmidae (Coleoptera, Dryopoidea)]. Acta Zootaxonomica Sinica 28 (2): 275–281 [in Chinese, with English summary].

Zhang, Z., C. Yang & D. Li, 1997. Coleoptera: Elmidae. In Wu, H. (ed.), Insects of Baishanzu Mountain, eastern China. [The Series of the Bioresources Expedition to the Baishanzu Mountain Nature Reserve]. China Forestry Publishing House, Beijing [1995]: 229–231 [in Chinese and English; publication date according to L. Ji (pers. comm.)].

Zhang, Z., C. Yang & L. Zhang, 2003. [English title: Five new species of Stenelmis (Coleoptera: Dryopoidea: Elmididae)]. Entomotaxonomia 25 (2): 118–124 [in Chinese and English].

Zimmermann, C., 1869. Stenelmis linearis, Stenelmis vittipennis, p. 259. In Le Conte, J.L. (ed.), Synonymical notes on Coleoptera of the United States, with descriptions of new species, from the MSS. of the late Dr. C. Zimmermann. Transactions of the American Entomological Society, II [1968–1969]: 243–259.

Zimmermann, L., 1908. Beiträge zur Kenntnis der mitteleuropäischen Dryopiden. Münchener Koleopterologische Zeitschrift 3 [1906–1908]: 341–345.

Index

Synonyms and unavailable names are italicized; a semicolon (";") is placed between unavailable names and their authors. Epithet names between square brackets refer to gender of original combination.

Family group names

Genus group names

Species group names